Environmental Health Criteria 222

BIOMARKERS IN RISK ASSESSMENT: VALIDITY AND VALIDATION

Published under the joint sponsorship of the United Nations Environment Programme, the International Labour Organization, and the World Health Organization, and produced within the framework of the Inter-Organization Programme for the Sound Management of Chemicals.

World Health Organization
Geneva, 2001

The **International Programme on Chemical Safety (IPCS)**, established in 1980, is a joint venture of the United Nations Environment Programme (UNEP), the International Labour Organization (ILO), and the World Health Organization (WHO). The overall objectives of the IPCS are to establish the scientific basis for assessment of the risk to human health and the environment from exposure to chemicals, through international peer-review processes, as a prerequisite for the promotion of chemical safety, and to provide technical assistance in strengthening national capacities for the sound management of chemicals.

The **Inter-Organization Programme for the Sound Management of Chemicals (IOMC)** was established in 1995 by UNEP, ILO, the Food and Agriculture Organization of the United Nations, WHO, the United Nations Industrial Development Organization, the United Nations Institute for Training and Research, and the Organisation for Economic Co-operation and Development (Participating Organizations), following recommendations made by the 1992 UN Conference on Environment and Development to strengthen cooperation and increase coordination in the field of chemical safety. The purpose of the IOMC is to promote coordination of the policies and activities pursued by the Participating Organizations, jointly or separately, to achieve the sound management of chemicals in relation to human health and the environment.

WHO Library Cataloguing-in-Publication Data

Biomarkers in risk assessment: validity and validation.

(Environmental health criteria ; 222)

1.Biological markers 2.Risk assessment – methods 3.Validation studies 4.Reproducibility of results 5.Environmental monitoring I.International Programme on Chemical Safety II.Series

ISBN 92 4 157222 1   (NLM Classification: QH 438.4.B55)
ISSN 0250-863X

The World Health Organization welcomes requests for permission to reproduce or translate its publications, in part or in full. Applications and enquiries should be addressed to the Office of Publications, World Health Organization, Geneva, Switzerland, which will be glad to provide the latest information on any changes made to the text, plans for new editions, and reprints and translations already available.

The designations employed and the presentation of the material in this publication do not imply the expression of any opinion whatsoever on the part of the Secretariat of the World Health Organization concerning the legal status of any country, territory, city or area or of its authorities, or concerning the delimitation of its frontiers or boundaries.

The mention of specific companies or of certain manufacturers' products does not imply that they are endorsed or recommended by the World Health Organization in preference to others of a similar nature that are not mentioned. Errors and omissions excepted, the names of proprietary products are distinguished by initial capital letters.

Computer typesetting by I. Xavier Lourduraj, Chennai, India

Printed in Finland
2001/14047 – Vammala – 5000

CONTENTS

ENVIRONMENTAL HEALTH CRITERIA FOR
BIOMARKERS IN RISK ASSESSMENT: VALIDITY
AND VALIDATION

NOTE TO READERS OF THE CRITERIA MONOGRAPHS

Every effort has been made to present information in the criteria monographs as accurately as possible without unduly delaying their publication. In the interest of all users of the Environmental Health Criteria monographs, readers are requested to communicate any errors that may have occurred to the Director of the International Programme on Chemical Safety, World Health Organization, Geneva, Switzerland, in order that they may be included in corrigenda.

* * *

A detailed data profile and a legal file can be obtained from the International Register of Potentially Toxic Chemicals, Case postale 356, 1219 Châtelaine, Geneva, Switzerland (telephone no. + 41 22 - 9799111, fax no. + 41 22 - 7973460, E-mail irptc@unep.ch).

* * *

This publication was made possible by grant number 5 U01 ES02617-15 from the National Institute of Environmental Health Sciences, National Institutes of Health, USA, and by financial support from the European Commission.

Environmental Health Criteria

PREAMBLE

Objectives

In 1973 the WHO Environmental Health Criteria Programme was initiated with the following objectives:

(i)   to assess information on the relationship between exposure to environmental pollutants and human health, and to provide guidelines for setting exposure limits;

(ii)   to identify new or potential pollutants;

(iii) to identify gaps in knowledge concerning the health effects of pollutants;

(iv)   to promote the harmonization of toxicological and epidemiological methods in order to have internationally comparable results.

The first Environmental Health Criteria (EHC) monograph, on mercury, was published in 1976 and since that time an ever-increasing number of assessments of chemicals and of physical effects have been produced. In addition, many EHC monographs have been devoted to evaluating toxicological methodology, e.g. for genetic, neurotoxic, teratogenic and nephrotoxic effects. Other publications have been concerned with epidemiological guidelines, evaluation of short-term tests for carcinogens, biomarkers, effects on the elderly and so forth.

Since its inauguration the EHC Programme has widened its scope, and the importance of environmental effects, in addition to health effects, has been increasingly emphasized in the total evaluation of chemicals.

The original impetus for the Programme came from World Health Assembly resolutions and the recommendations of the 1972 UN Conference on the Human Environment. Subsequently the work became an integral part of the International Programme on Chemical Safety (IPCS), a cooperative programme of UNEP, ILO and WHO.

In this manner, with the strong support of the new partners, the importance of occupational health and environmental effects was fully recognized. The EHC monographs have become widely established, used and recognized throughout the world.

The recommendations of the 1992 UN Conference on Environment and Development and the subsequent establishment of the Intergovernmental Forum on Chemical Safety with the priorities for action in the six programme areas of Chapter 19, Agenda 21, all lend further weight to the need for EHC assessments of the risks of chemicals.

Scope

The criteria monographs are intended to provide critical reviews on the effect on human health and the environment of chemicals and of combinations of chemicals and physical and biological agents. As such, they include and review studies that are of direct relevance for the evaluation. However, they do not describe *every* study carried out. Worldwide data are used and are quoted from original studies, not from abstracts or reviews. Both published and unpublished reports are considered and it is incumbent on the authors to assess all the articles cited in the references. Preference is always given to published data. Unpublished data are used only when relevant published data are absent or when they are pivotal to the risk assessment. A detailed policy statement is available that describes the procedures used for unpublished proprietary data so that this information can be used in the evaluation without compromising its confidential nature (WHO (1999) Guidelines for the Preparation of Environmental Health Criteria. PCS/99.9, Geneva, World Health Organization).

In the evaluation of human health risks, sound human data, whenever available, are preferred to animal data. Animal and *in vitro* studies provide support and are used mainly to supply evidence missing from human studies. It is mandatory that research on human subjects is conducted in full accord with ethical principles, including the provisions of the Helsinki Declaration.

The EHC monographs are intended to assist national and international authorities in making risk assessments and subsequent risk management decisions. They represent a thorough evaluation of risks and are not, in any sense, recommendations for regulation or standard setting. These latter are the exclusive purview of national and regional governments.

Content

The layout of EHC monographs for chemicals is outlined below.

* Summary – a review of the salient facts and the risk evaluation of the chemical
* Identity – physical and chemical properties, analytical methods
* Sources of exposure
* Environmental transport, distribution and transformation
* Environmental levels and human exposure
* Kinetics and metabolism in laboratory animals and humans
* Effects on laboratory mammals and *in vitro* test systems
* Effects on humans
* Effects on other organisms in the laboratory and field
* Evaluation of human health risks and effects on the environment
* Conclusions and recommendations for protection of human health and the environment
* Further research
* Previous evaluations by international bodies, e.g. IARC, JECFA, JMPR

Selection of chemicals

Since the inception of the EHC Programme, the IPCS has organized meetings of scientists to establish lists of priority chemicals for subsequent evaluation. Such meetings have been held in Ispra, Italy, 1980; Oxford, United Kingdom, 1984; Berlin, Germany, 1987; and North Carolina, USA, 1995. The selection of chemicals has been based on the following criteria: the existence of scientific evidence that the substance presents a hazard to human health and/or the environment; the possible use, persistence, accumulation or degradation of the substance shows that there may be significant human or environmental exposure; the size and nature of populations at risk (both human and other species) and risks for environment; international concern, i.e. the substance is of major interest to several countries; adequate data on the hazards are available.

If an EHC monograph is proposed for a chemical not on the priority list, the IPCS Secretariat consults with the Cooperating Organizations and all the Participating Institutions before embarking on the preparation of the monograph.

Procedures

The order of procedures that result in the publication of an EHC monograph is shown in the flow chart on p. x. A designated staff member of IPCS, responsible for the scientific quality of the document, serves as Responsible Officer (RO). The IPCS Editor is responsible for layout and language. The first draft, prepared by consultants or, more usually, staff from an IPCS Participating Institution, is based initially on data provided from the International Register of Potentially Toxic Chemicals, and reference data bases such as Medline and Toxline.

The draft document, when received by the RO, may require an initial review by a small panel of experts to determine its scientific quality and objectivity. Once the RO finds the document acceptable as a first draft, it is distributed, in its unedited form, to well over 150 EHC contact points throughout the world who are asked to comment on its completeness and accuracy and, where necessary, provide additional material. The contact points, usually designated by governments, may be Participating Institutions, IPCS Focal Points, or individual scientists known for their particular expertise. Generally some four months are allowed before the comments are considered by the RO and author(s). A second draft incorporating comments received and approved by the Director, IPCS, is then distributed to Task Group members, who carry out the peer review, at least six weeks before their meeting.

The Task Group members serve as individual scientists, not as representatives of any organization, government or industry. Their function is to evaluate the accuracy, significance and relevance of the information in the document and to assess the health and environmental risks from exposure to the chemical. A summary and recommendations for further research and improved safety aspects are also required. The composition of the Task Group is dictated by the range of expertise required for the subject of the meeting and by the need for a balanced geographical distribution.

EHC PREPARATION FLOW CHART

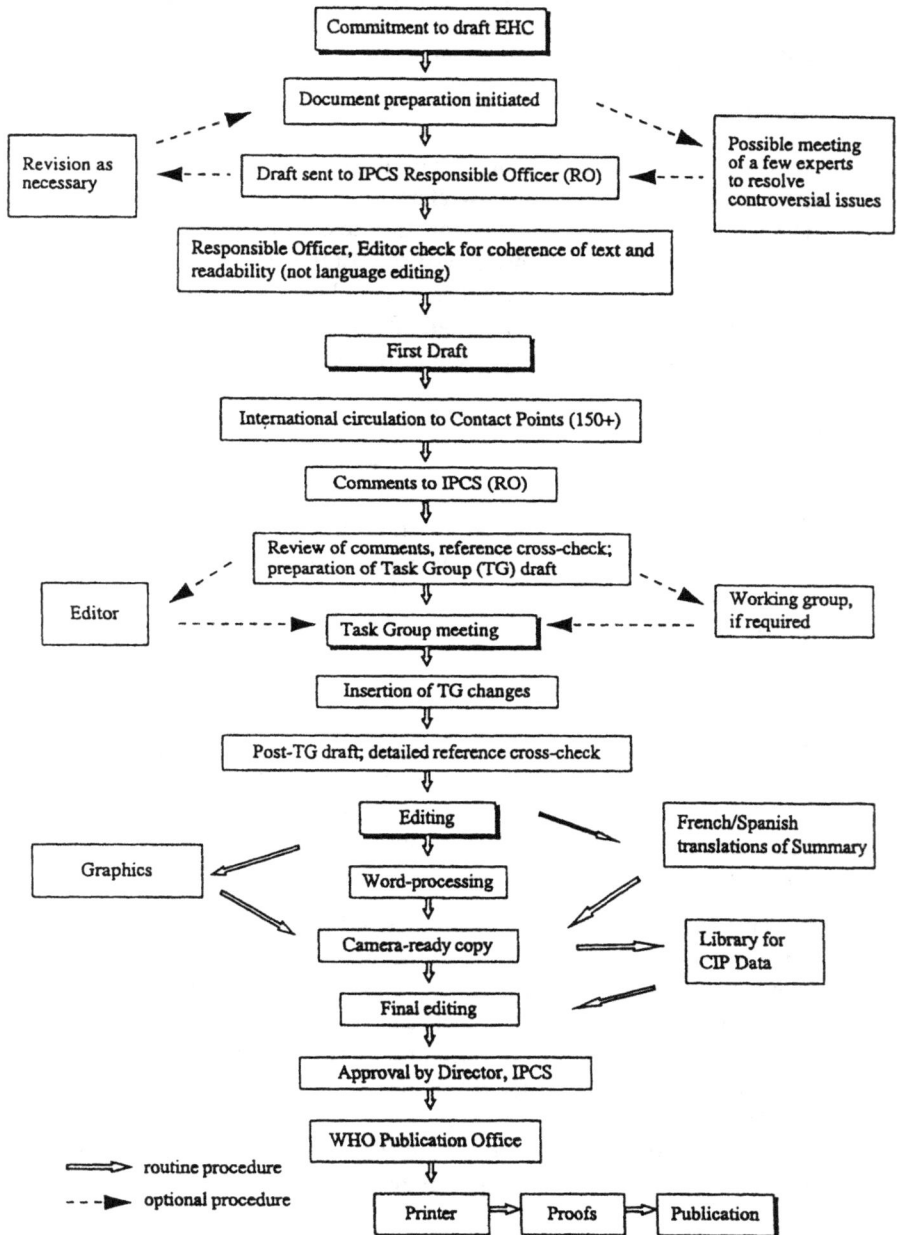

| Commitment to draft EHC |

⇩

| Document preparation initiated |

⇩

| Revision as necessary | ◀- - - | Draft sent to IPCS Responsible Officer (RO) | ◀- - - | Possible meeting of a few experts to resolve controversial issues |

⇩

| Responsible Officer, Editor check for coherence of text and readability (not language editing) |

⇩

| First Draft |

⇩

| International circulation to Contact Points (150+) |

⇩

| Comments to IPCS (RO) |

⇩

| Review of comments, reference cross-check; preparation of Task Group (TG) draft |

⇩

| Editor | - - - ▶ | Task Group meeting | ◀- - - - - | Working group, if required |

⇩

| Insertion of TG changes |

⇩

| Post-TG draft; detailed reference cross-check |

⇩

| Editing | → | French/Spanish translations of Summary |

⇩

| Graphics | ◀ | Word-processing |

⇩

| Camera-ready copy | → | Library for CIP Data |

⇩

| Final editing |

⇩

| Approval by Director, IPCS |

⇩

| WHO Publication Office |

⇩

⟹ routine procedure
- - ▶ optional procedure

| Printer | ⇨ | Proofs | ⇨ | Publication |

The three cooperating organizations of the IPCS recognize the important role played by nongovernmental organizations. Representatives from relevant national and international associations may be invited to join the Task Group as observers. Although observers may provide a valuable contribution to the process, they can only speak at the invitation of the Chairperson. Observers do not participate in the final evaluation of the chemical; this is the sole responsibility of the Task Group members. When the Task Group considers it to be appropriate, it may meet *in camera*.

All individuals who as authors, consultants or advisers participate in the preparation of the EHC monograph must, in addition to serving in their personal capacity as scientists, inform the RO if at any time a conflict of interest, whether actual or potential, could be perceived in their work. They are required to sign a conflict of interest statement. Such a procedure ensures the transparency and probity of the process.

When the Task Group has completed its review and the RO is satisfied as to the scientific correctness and completeness of the document, it then goes for language editing, reference checking and preparation of camera-ready copy. After approval by the Director, IPCS, the monograph is submitted to the WHO Office of Publications for printing. At this time a copy of the final draft is sent to the Chairperson and Rapporteur of the Task Group to check for any errors.

It is accepted that the following criteria should initiate the updating of an EHC monograph: new data are available that would substantially change the evaluation; there is public concern for health or environmental effects of the agent because of greater exposure; an appreciable time period has elapsed since the last evaluation.

All Participating Institutions are informed, through the EHC progress report, of the authors and institutions proposed for the drafting of the documents. A comprehensive file of all comments received on drafts of each EHC monograph is maintained and is available on request. The Chairpersons of Task Groups are briefed before each meeting on their role and responsibility in ensuring that these rules are followed.

WHO TASK GROUP ON ENVIRONMENTAL HEALTH CRITERIA FOR BIOMARKERS IN RISK ASSESSMENT: VALIDITY AND VALIDATION

Members

Dr D. Anderson, TNO BIBRA International Ltd, Carshalton, Surrey, United Kingdom (*Rapporteur*)

Dr H. Autrup, Department of Environmental Medicine, University of Aarhus, Aarhus, Denmark (*Chairman*)

Dr S. Bonassi, Department of Environmental Epidemiology, National Institute for Research on Cancer, Genoa, Italy

Dr K. Hemminki, Department of Biosciences at Novum, Karolinska Institute, Huddinge, Sweden

Dr A. Mutti, Laboratory of Industrial Toxicology, Department of Clinical Medicine, Nephrology, and Health Sciences, University of Parma Medical School, Parma, Italy

Dr O. Pelkonen, Department of Pharmacology and Toxicology, University of Oulu, Oulu, Finland

Dr P.A. Schulte, Education and Information Division, National Institute for Occupational Safety and Health, Cincinnati, Ohio, USA

Secretariat

Dr A. Aitio, International Programme on Chemical Safety, World Health Organization, Geneva, Switzerland (*Joint Secretary*)

Dr Y. Hayashi, International Programme on Chemical Safety, World Health Organization, Geneva, Switzerland (*Joint Secretary*)

WHO TASK GROUP ON ENVIRONMENTAL HEALTH CRITERIA FOR BIOMARKERS IN RISK ASSESSMENT: VALIDITY AND VALIDATION

A WHO Task Group on Environmental Health Criteria for Biomarkers in Risk Assessment: Validity and Validation met at TNO BIBRA International, Carshalton, Surrey, United Kingdom from 3 to 6 April 2000. Dr A. Aitio, IPCS, welcomed the participants on behalf of the IPCS and its three cooperating organizations (UNEP/ILO/WHO). The Task Group reviewed and revised the draft monograph.

This Environmental Health Criteria monograph is composed of the main text and four authored papers. The main text was constructed by Dr P.A. Schulte, based on the source documents and was reviewed by the IPCS Contact Points. The comments received were considered by the principal author, and the revisions were discussed and approved by the Task Group. The source documents were similarly subjected to IPCS review and were then revised accordingly by the authors. However, they were not discussed thoroughly during the Task Group meeting and thus represent the views of the authors.

Dr A. Aitio and Mr Y. Hayashi of the IPCS Central Unit were responsible for the overall scientific content of the monograph and Dr P.G. Jenkins of the IPCS Central Unit was responsible for the technical editing of the monograph.

The efforts of all who helped in the preparation of the monograph are gratefully acknowledged.

* * *

The preparation of the draft was financially supported by the US Environmental Agency. Financial support for this Task Group was provided by the UK Department of Health as part of its contribution to the IPCS.

ABBREVIATIONS

BMD	benchmark dose
BP	benzo(a)pyrene
CYP	cytochrome P450
GC	gas chromatography
GST	glutathione *S*-transferase
IARC	International Agency for Research on Cancer
LOAEL	lowest-observed-adverse-effect level
MS	mass spectroscopy
NAT	*N*-acetyltransferase
NOAEL	no-observed-adverse-effect level
PAH	polycyclic aromatic hydrocarbon
PCR	polymerase chain reaction
PM	poor metabolizer
TCDD	tetrachlorinated dibenzo-*p*-dioxin
TLV	threshold limit value
TWA	time-weighted average
XME	xenobiotic metabolizing enzyme

1. INTRODUCTION

The aim of risk assessments is to provide society with estimates of the likelihood of illnesses and injury as a consequence of exposure to various hazards. Risk assessments are needed when social policy decisions are in dispute, when the health consequences of alternative policies in question are not subject to direct measurement (at least in a timely fashion), and when the scientific analysis of a hazard is not complete (Hattis & Silver, 1993). The assessment procedure involves the development of an exposure-response curve for the target species (e.g., humans), based on animal and human information, followed by the projection of the curves to estimate levels of exposure that may be considered safe (NRC, 1987). For risk assessments to be useful they should lead to projections that are close to the true risks. A strong scientific basis for conducting risk assessments is the best way to assure that projections are close to true risks or at least provide an honest depiction of the state of knowledge and the degree of certainty about risks (Bailar & Bailer, 1999).

Risk assessment has a range of meanings. At the basic level it is an exercise to evaluate the potential of some hazard to induce an adverse human health response. It can be a qualitative or quantitative exercise at the individual or group (population) level. The term quantitative risk assessment (QRA) has been used to describe the response associated with a specific level of exposure (Bailer & Dankovic, 1997). The availability of adequate dose/concentration-response data is a prerequisite to conducting a QRA.

A biomarker is any substance, structure or process that can be measured in the body or its products and influence or predict the incidence of outcome or disease. Biomarkers can be classified into markers of exposure, effect and susceptibility. If biomarkers are to contribute to environmental and occupational health risk assessments, they have to be relevant and valid. Relevance refers to the appropriateness of biomarkers to provide information on questions of interest and importance to public and environmental health authorities and other decision-makers. The use of relevant biomarkers allows decision-makers to answer important public health questions by being used in research or risk assessments in a

1

way that contributes useful information that cannot be obtained better by other approaches, such as questionnaires, environmental measurements or record reviews. For example, chronic exposure to organochlorines is better indicated by serum organochlorine levels than by market-basket studies or industrial hygiene measurements, and early kidney damage may be better indicated by a battery of urinary biomarkers than by morbidity records. Relevance also pertains to whether the questions on which a biomarker can provide information are important questions; not merely ones that can be answered, but ones that should be answered (Muscat, 1996). Thus, the ability to measure a biomarker after exposure to a toxicant may not be as important a question as whether individuals with exposure to the toxicant are at increased risk of disease.

The second characteristic of potentially useful biomarkers is validity. Validity of biomarkers has been widely discussed (Hernberg & Aitio, 1987; Schatzkin et al., 1990; Schulte & Perera, 1993; Boffetta, 1995; Bernard, 1995; Dor et al., 1999). It includes both laboratory and epidemiological aspects. Validity refers to a range of characteristics that is the best approximation of the truth or falsehood of a biomarker. It is a sense of degree rather than an all-or-none state. The validity of a biomarker is a function of intrinsic qualities of the biomarker and characteristics of the analytic procedures (Dor et al., 1999) (see Tables 1 and 2 for an example of this distinction). Additionally, three broad categories of validity can be distinguished: measurement validity, internal study validity and external validity (Schulte & Perera, 1993). Measurement validity (in terms of analytical chemistry, accuracy) is the degree to which a biomarker indicates what it purports to indicate. Internal study validity is the degree to which inferences drawn from a study actually pertain to study subjects and are true. External validity is the extent to which findings of a study can be generalized to apply to other populations. The use of invalid biomarkers can lead to invalid inferences and generalizations and ultimately to erroneous risk assessments.

Although biomarkers have a long history in medicine and public health, the systematic development, validation and application of biomarkers is a relatively new field in environmental health (Shugart et al., 1992; Anderson S et al., 1994), except for biological monitoring in occupational health (Hernberg & Aitio, 1987).

Table 1. Factors affecting the validity and feasibility of
biomarker studies: analytical procedures

- Sampling constraints (for example, timing requirements)

- Number of samples necessary for an acceptable precision

- Degree of invasiveness of the sampling procedure

- Availability of storage methods after the sample is taken (to avoid the
  need for immediate analysis)

- Controlling or reducing the contamination of the sample when it is
  taken and when it is manipulated in the laboratory

- Simplicity, possibility of routine usage, and speed of the procedure

- Trueness, precision and sensitivity

- Specificity for the component to be detected: interference must be
  identified to avoid misinterpretation

- Standardization of the procedure

Adapted from: Dor et al. (1999)

Table 2. Factors affecting the validity of biomarkers:
intrinsic characteristics of the biomarker

- Significance: exposure, effect, individual susceptibility

- Specificity in relation to the pollutant or pollutant family

- Sensitivity: capacity to distinguish populations with different exposure
  levels, susceptibilities or degrees of effect

- Knowledge of its background in the general population

- Existence of dose-response curves between exposure level and
  marker concentration

- Estimation of the inter- and intra-individual variability

- Knowledge of confounding factors that can affect marker

Adapted from: Dor et al. (1990)

3

However, such efforts are not new in others areas, e.g., in the validation of serum lipid biomarkers in cardiovascular disease. Lessons can be learned from the cardiovascular field that can be applied to the environmental health field; notably that the validation of markers for risk assessment can take a long time and is generally expensive. In the validation efforts, laboratory scientists and epidemiologists, clinicians, exposure assessors and statisticians need to be involved. In addressing societal impediments to the validation, an even broader range of disciplines, such as ethics, laws, and economics also need to participate.

When used in risk assessment, information from biological markers may replace default assumptions when specific information regarding exposure, absorption and toxicokinetics is unavailable or limited (Table 3) (Ponce et al., 1998). Although examples of how this biomarker information can be used are limited, a general framework can be adduced (see section 2.2). Quantitative evaluations of the utility of biomarker information in risk assessment are rare (Bois et al., 1995; Ponce et al., 1998).

One compelling example of the use of susceptibility markers is the work of El-Masri et al. (1999). They investigated how changing glutathione-S-transferase theta (GSTT1) genotype frequencies would impact cancer risk estimates from dichloromethane by the application of Monte-Carlo simulation methods in combination with physiologically based pharmacokinetic (PBPK) models. They reported that average and median risk estimates were 23% to 30% higher when GSTT1 polymorphism was not included in the models. This analysis was a major factor in the permissible exposure levels promulgated by the US Occupational Safety and Health Administration (OSHA, 1998).

Goldstein (1996) has identified two important impediments to the development of biomarkers of value to risk assessment. The first is the over-reliance on mathematical models to the exclusion of monitoring data. This occurs because regulators have a need to make a decision and, for expedience, use models until better approaches are developed. However, once locked into a regulation, the existence of the model serves as a major inhibition to the development of more reliable methods of indicating exposure and effect, including biomarkers. The second is that ethical review

Table 3. Use of biomarkers to refine risk assessment information

Variable	Use of biomarkers
Exposure	Establish exposure characteristics • Route of exposure • Peak of exposure • Total exposure Estimate cumulative exposure
Absorption	Establish absorption factors • Inhalation • Dermal exposure • Ingestion Identify factors that influence absorption Identify interspecies differences Identify sensitive population characteristics
Toxicokinetics	Establish distribution kinetics Establish half-life in blood or body Identify interspecies differences Identify factors that influence distribution, metabolism or excretion Estimate cumulative exposure Estimate peak exposure variables • Time • Concentration Identify sensitive population characteristics
Toxicodynamics	Identify mechanism of toxicity at target organ Establish target organ potency Identify sensitive population characteristics Identify factors that influence target organ toxicity Identify interspecies differences

(Ponce et al., 1998)

boards may find it difficult to sanction research where participants are exposed in a scientific study to levels they are exposed to in the general environment or at work because the participation in the study is voluntary while the latter is generally involuntary.

The concepts and principles supporting the use of biomarkers in the assessment of human health risks from exposure to chemicals have been reviewed by the International Programme on Chemical Safety (IPCS, 1993). The IPCS has produced the concise guidelines for the monitoring of genotoxic effects of carcinogens in humans (Albertini et al., 2000). It has also issued monographs on the methodology for the assessment of human health risks in a wider context, which includes the use of biomarkers (IPCS, 1994, 1999).

2. RISK ASSESSMENT

The widely accepted risk assessment paradigm includes the steps of hazard identification, dose-response assessment, exposure assessment and risk characterization (Fig. 1) (NRC, 1983). Using biomarkers to gauge exposure may contribute in various steps of the risk assessment process. In the hazard identification step, i.e., the determination of whether an agent might pose a threat to human health, there is a need to link an exposure with an adverse outcome. Given the different effects at different levels of exposure, there is need for understanding the specific effects of different exposures, particularly at lower levels of exposure. Then, in the exposure assessment stage, the extent of exposure is highly dependent on the agent and environment and builds on the specific source-path-receiver model utilized during hazard identification. The source-path-receiver model is the common approach to link source chemicals, the pathway of movement in the environment, and the route(s) of exposure of various receptors, in the case of risk assessment, individuals or groups of individuals (Nelson, 1997). Critical issues in exposure assessment include characterization of the magnitude, frequency and duration of exposure, the basis for the assessment and the identification of highly exposed subgroups. The risk characterization step requires consideration of any assumptions and models used, and attendant uncertainties used in developing the risk estimates. These estimates are then the basis for options to be selected in the risk management stage (Schulte & Waters, 1999).

Quantitative estimation of health risks is dependent on both exposure characterization and the nature of the dose-response relationships or toxicity of the agents involved. The greatest uncertainties in risk assessment almost always arise from sparse or inadequate exposure data, inadequate understanding of mechanisms of toxicity, and insufficient understanding of the exposure-dose-response pathway (Becking, 1995; McClellan, 1995). Two additional factors can lead to uncertainties in risk assessments. These include mixed or multiple exposures implicated in the disease pathway, and variability of both exposures and responses within and between individuals.

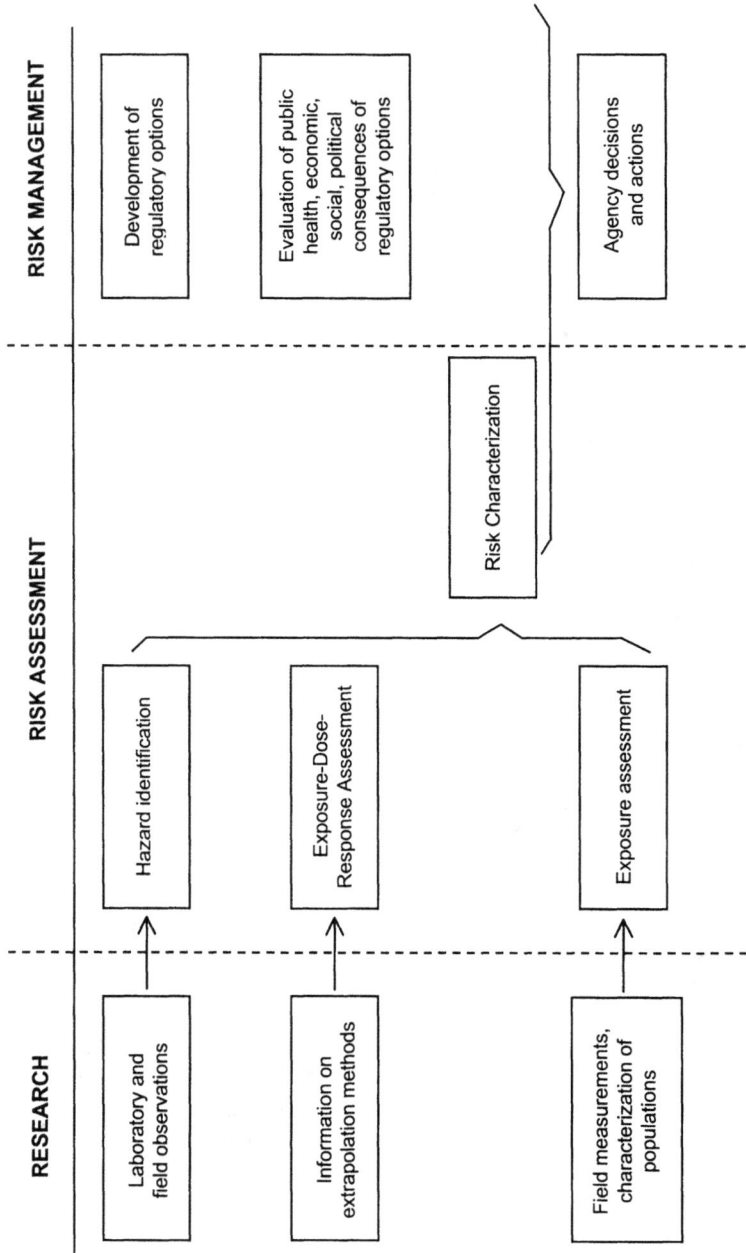

Fig. 1. National Academy of Sciences/National Research Council Paradigm for Research/Risk Assessment/Risk Management (adapted from: NAS/NRC, 1983)

There are ambiguities in the risk assessment terminology that should be identified. For example, consideration of exposure will occur in two places in the risk assessment model. In the hazard identification stage, exposure is a component of the underlying research. This is distinct from the exposure assessment stage of risk assessment, where the extent of exposure of the population (whose risk is being characterized) to the identified hazard is determined. In a similar sense, dose-response considerations appear in two places. One of the criteria for the identification of a hazard is the finding of a dose-response relationship in the component studies. Additionally in the dose-response stage of risk assessment, the objective is to ascertain if there is a dose-response relationship in all the available data, identifying the shape of the curve and projecting to exposure or dose level, where health effects are reduced or believed absent. Finally, the concept of susceptibility can be in action throughout the risk assessment model. In hazard identification gene-environment interaction or effect modification may be assessed, and similarly in the dose-response stage susceptibility may be taken into account. Finally, in the risk characterization stage, different risk projections could be determined for various population subgroups identified by susceptibility factors.

2.1 Hazard identification

Historically, hazard identification has been the driving force in risk assessment (McClellan, 1999). Various national and international organizations have recognized the role human biological markers of exposure and effect can play in the hazard identification step: both make use of such data in classifying carcinogens.

Like other classic measures of exposure, there are limitations to the use of exposure biomarkers in epidemiological research for hazard identification (Schulte & Perera., 1993; Pearce et al., 1995). The major limitation is the general inability of biomarkers (with some exceptions) to indicate historic exposures. Additionally, their strength in integrating routes of exposure also may be a weakness by introducing confounding due to source, as full use of exposure biomarkers may also require understanding of those inherited and acquired factors that influence the level of exposure biomarkers (Vineis et al., 1990).

The role of biomarkers in hazard identification can be considered in the following examples. In determining whether a xenobiotic is hazardous, biomarkers may yield a more accurate determination than approaches based on less sensitive measures of exposures (e.g., job titles, as exposure proxies). In situations where exposures occurred that were variable or intermittent and the effect of exposure is integrated, biomarkers that represent a cumulation of exposure, such as haemoglobin adducts, might be useful (Perera, 1995). A biomarker approach may allow for clarifying exposure-outcome relationships better than with classical methods, due to reduction in exposure measurement error. For example, the role of aflatoxin exposure and liver cancer was not clear when studied by dietary questionnaire to assess intake of foods potentially contaminated with aflatoxin. However, a strong association was observed based on urinary biomarkers (metabolites and nucleic acid adducts) of aflatoxin exposure (Qian et al., 1994; Howe, 1998) (Table 4). Usually information is not available to permit this kind of comparison. In this example the molecular biomarker is useful because it is quite specific and biologically relevant, and it provides a better indicator of exposure than could be inferred from a questionnaire since respondents are not aware of how much aflatoxin they consume. One might ask if direct measurement of the aflatoxin component of all foodstuffs ingested and measurement of amounts of food intake per day might also lead to a better measure of exposure than the questionnaire surrogate. In the case of aflatoxin this is probably not true due to the difficulty of measuring food intake, the possible variability of aflatoxin levels within food, the difficulty of extractions of aflatoxin from foods, and analytical detection limits for such methods. However, for some other agents, external direct measures of exposures may be both feasible and as cost effective as biological measures, and will also provide improved estimation over such surrogates as questionnaire data.

The basic rationale for using exposure biomarkers is that they could provide, in some cases, a more accurate method for assessing exposure and, ultimately, risk (Fig. 2) (Schulte & Waters, 1999). While use of biomarkers can reduce mis-classification, it is also possible that measurement error in the biomarker may contribute to bias in the measure of association (White, 1997; Saracci, 1997). Such error can be evaluated and its impact adjusted for, but, on balance, it is better to avoid or minimize it with good laboratory and epidemiological practices.

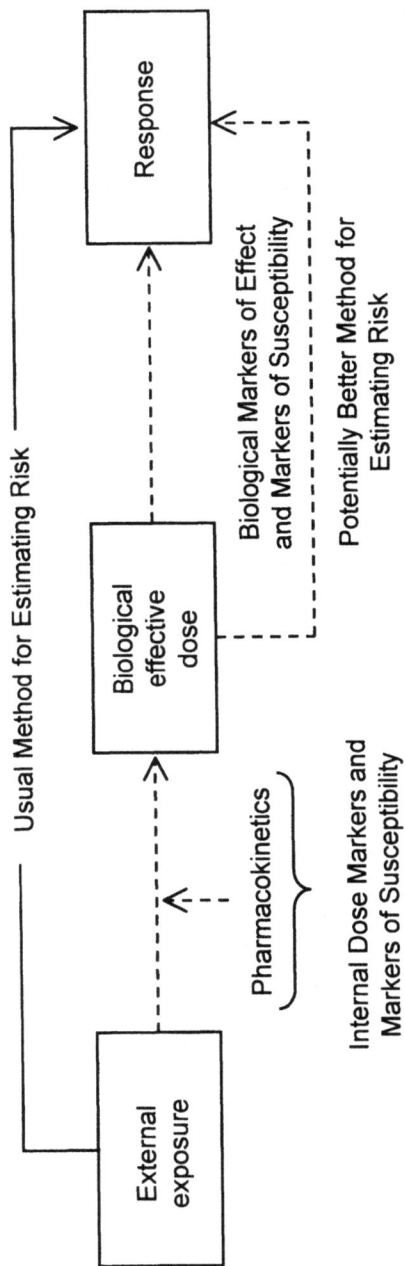

Fig. 2. Rationale for using biomarkers to assess risk
(Schute & Waters, 1999)

11

Table 4. Comparison between estimated dietary intake of aflatoxin and biomarker derived exposure data

A. Relative risks based on estimated dietary intake of aflatoxin[a]

Dietary aflatoxin B_1 exposure (µg/year)	Relative risk	95% Confidence interval
< 71	1.0	
71–113	1.6	0.8 – 3.1
113+	0.9	0.4 – 1.9

[a] Estimate of the intake of aflatoxin was based on a dietary questionnaire to assess the intake of foods potentially contaminated by aflatoxins and analysis of related food items for aflatoxin contents.

B. Relative risks based on urinary biomarkers of aflatoxin

Biomarker	Relative risk (present/absent)	95% Confidence interval
Metabolites or adducts	5.0	2.1 – 12
Adducts only	16	3.6 – 73

Source: Study of aflatoxin and risk of liver cancer in Shanghai, China from 1986 to 1992
(No. of cases = 55) (Qian et al., 1994; adapted from Howe, 1998)

The value of biological markers in epidemiological studies that will be useful in the hazard identification step of risk assessment depends on the quality of the design and analysis of the studies. Various reviews of biomonitoring and molecular epidemiological studies have been conducted and in some instances have found that a better design or analysis could have been applied (Bonassi et al., 1994; IARC, 1997). In 1994, Bonassi et al. reviewed three years of biomonitoring studies and found that only 5% of those studies adopted the best statistical techniques available. The major recommendations were to focus more on point estimates and confidence intervals instead of significant tests, utilize appropriate multivariate techniques and pay more attention to adjustment for confounding and evaluation of possible interaction between factors.

2.2 Dose response

One of the most controversial aspects of risk assessment is the extrapolation of higher-level exposure data to lower levels of exposure (Goldstein, 1996). In risk assessment the ascertainment of a dose-response relationship is crucial for ultimately determining the shape of the dose-response curve and for predicting a no-observed-adverse-effect level (NOAEL). Exploring the lower end of the dose-response curve through epidemiological studies is generally impractical, if not impossible, due to extraordinarily large sample sizes (Stayner, 1992). However, biomarkers can contribute to identifying a dose-response relationship at lower levels of exposure. The demand on environmental epidemiology to evaluate increasingly subtle health risks requires more accurate estimation of the quantity and timing of a toxicant reaching target tissue (Kriebel, 1994). Kriebel (1994) described a two-stage approach to derive estimates of dose from exposure data and then linked them to epidemiological models estimating disease risk. Such an approach incorporates physiological processes into epidemiological modeling and is possibly more valid than approaches with less detail. Biomarkers of exposure can be used as indicators of dose, which can then be assessed against classic measures of morbidity or mortality.

Often dose-response determinations are made by use of PBPK models. Examples of how biomarker data could be incorporated in PBPK models include: 1) calibration of the model (empirically determining the population values of kinetic parameters); 2) validation of the model (determining how well the model predicts data of another cohort); 3) prediction (applying the model to new cohorts, and predicting the internal doses associated with given exposure scenarios). These predicted measures of internal (biologically-effective) dose can then be used in dose-response modelling, in lieu of the external exposure measures, to predict disease risk. The internal dose measures may be better predictors of disease risk, especially when exposure-dose is nonlinear (e.g., due to capacity-limited metabolism). A PBPK model could also potentially allow for extrapolation from limited data, such as a short-term laboratory study to predict the biomarker concentrations that might be found in a population. For example, the model for carboxyhaemoglobin formation (Andersen et al., 1991) from exposure to methylene chloride was calibrated against short-term human exposure but could be used to predict carboxyhaemoglobin from long-term exposures.

Another use of biomarkers can be as outcome measures that correlate with exposure. Here exposure markers are not what is needed; rather, there is a need for effect markers. Effect markers are those that relate to or predict disease. A marker validated to predict disease can be used as a surrogate for disease. For example, specific types of chromosomal aberrations that appear to predict cancer risk on a group basis could be used as the outcome variables in a dose-response analysis, as in the case of ionizing radiation exposure (Joksić & Spasojević-Tišma, 1998).

Persistent microalbuminuria and low molecular weight proteinuria identify incipient diabetic and cadmium-induced nephropathy, respectively (IPCS, 1992; Emancipator, 1999). Proteinuria is also the main factor accelerating progressive renal disease toward end-stage renal failure (for review, see Remuzzi & Bertani, 1998), which confers a prognostic value to its quantitative changes over upper reference limits.

The literature on biomarkers in PBPK models is relatively limited. If there are enough biomarker data there may be no need for the PBPK model at all. This would be particularly true for biomarkers of effect, if the human biomarker data came from the population that was at risk, or one with similar exposures.

2.3 Exposure assessment for risk assessment

Exposure assessment has generally been considered the weak link in hazard evaluation and risk assessment (Dary et al., 1996). The exposure assessment component of risk assessment includes consideration of such issues as representativeness of exposure measurements for a population, differences in exposures within and between individuals, individual differences in uptake and biotransformation, identification of factors that control or modify exposures, exposure estimation methods applicable in the absence of direct measurements, and identification of the most relevant dose metric (the most relevant measure of dose) for the agent under consideration (Schulte & Waters, 1999). The use of biomarkers in assessing exposure for risk assessment increasingly may include consideration of susceptibility factors in conjunction with exposure factors (gene-environment interaction), such as the presence of a specific genetic polymorphism for a metabolic enzyme (Bois et al., 1995; D'Errico et al., 1996). Such genetic differences may account for some interindividual variability in exposure markers.

Quite often epidemiological studies utilize exposure surrogates rather than direct measurement of exposures. For environmental studies, surrogates might include geographical location such as residence for a drinking-water or air pollution study, age of housing in studies of lead-based paint exposures, or proximity of residence to electrical power lines. Occupational studies use surrogates such as job title or job group, years worked at a plant, pounds of pesticide applied per week, and tasks performed when direct measurements are not available or are limited (Stewart et al., 1991; Goldberg & Hémon, 1993). The use of quantified direct measurements of personal exposures can lower uncertainty in the risk assessment process considerably compared to the use of such exposure surrogates (Schulte & Waters, 1999). Biomarkers may serve to evaluate the completeness of exposure assessment information by associating environmental or source information, exposure measurements, and epidemiological and human activity data with internal dose (Dary et al., 1996).

In some cases, biomarkers of exposure may be better than external measurements of exposure for situations where protective equipment has been used or when there is the possibility of dermal (or gastrointestinal) absorption.

3. VALIDITY AND VALIDATION –
GENERAL CONSIDERATIONS

The ultimate driving force for whether biomarkers will contribute to environmental health efforts is the validity of the markers. Validity is a complex characteristic that describes the extent to which a biomarker reflects a designated event in a biological system. Generally, these events are exposure, effects of exposure, disease and susceptibility. Validity has meaning according to discipline as well. To the laboratory scientist, validity often refers to the nature of the biomarker and the characteristic of the assay for the biomarker. Thus, the sensitivity of the assay to detect a signal at a given concentration, and the ability of the signal to be specific for a particular event are indications of validity to the laboratory scientist. In addition, the scientist wants to know what factors might influence an assay. The epidemiologist relies on the laboratory definition of validity as the cornerstone of population studies, but then needs to know how likely a person with a positive assay or test is to develop disease (or have been exposed) and how likely a person with a negative test is to be free of disease (or exposure). The epidemiologist also needs to know how feasible the marker is to use in human populations and the reliability of the assay under field conditions. Moreover, the epidemiologist needs to know how the frequency of the marker varies in different population subgroups defined by age, race, gender, pre-existing illness, diet and various behavioral factors. Only when validity at the laboratory and population level has been established is a biomarker ready for the full spectrum of environmental research and uses. As noted, most biomarkers have not had that level of validation. A broad effort is underway but the products of this activity are not available yet.

Validation of candidate biomarkers is an empirical process that can be approached by producing several different, but convergent lines of evidence. There is extensive literature on criteria for validating biological markers (e.g., WHO, 1975; Lucier & Thompson, 1987; Hernberg & Aitio, 1987; Schulte, 1989; Schatzkin et al., 1990; Margetts, 1991; Schulte & Mazzuckelli, 1991; Schulte & Perera, 1993; Schulte & Talaska, 1995; Boffetta, 1995; Ponce et al., 1998). In general, these criteria include understanding the natural history, biological and temporal relevance, pharmacokinetics, background variability, dose response and confounding factors (Schulte & Talaska, 1995). Biomarker validity is also dependent on reliability of the assay to measure the biomarkers. These criteria allow for the assessment of whether a biomarker represents an event that is in a continuum between exposure and resultant disease, and whether the biological specimen containing the biomarkers is appropriate and the marker reflects the time period of concern. Finally, by assessing confounding and effect modifying factors, it is possible to understand what other factors influence a biomarker or its assay.

The careful measurement of strong confounders and effect modifiers should be given as much attention as is given to measurement of the exposure and disease variables or biomarkers. Consideration should be given to mounting validation substudies to quantify measurement error in important covariates (Hatch & Thomas, 1993). Measurements of biological markers are the building blocks of research and mechanistically based risk assessment. If the measurements are inaccurate, the research and risk assessments are likely also to be biased. Controlling measurement validity makes it possible partially to control study validity since measurement errors can produce biased estimates of regression coefficients used in statistical models of exposure and disease (Louis, 1988). Measures of association, such as the odds ratio, can be distorted, depending on the type of error and other characteristics, towards or away from the null hypotheses of no association between the biomarker and disease (or exposure).

The terminology to express measurement error traditionally used in biomarker measurements is different from that applied in analytical chemical laboratories. In the latter, traditionally, variability of the results is considered on an individual basis, and accuracy (trueness and precision) refers to individual analytical results. Trueness refers to how close to the true value the average of the results is, precision to the scatter of the results around their average, while accuracy is the combination of the two characteristics. Defective trueness is bias, defective precision, imprecision. In epidemiological work involving biomarkers, the emphasis is on the biomarker level in the studied population rather than on an individual, and the measurement error includes the intra-individual variation with time (White, 1997). As White (1997) notes:

measurement errors for an individual can be defined as the difference between a person's measured biomarker (the biomarker "test") and the person's true biomarker. The true biomarker is the biomarker without laboratory or other sources of error and, if the measure can fluctuate over time, the true biomarker is an integration of its concentration over the time period of etiological interest.

Validity in the context of epidemiological research involving biomarkers can be defined as the relation of the biomarker test (the potentially mismeasured biomarker) to true biomarker in the population of interest. Parameters that describe the measurement error in the population are called measures of validity (White, 1997). Two indicators of measurement error are used to describe the validity of an observed measurement compared with the true measurement (Armstrong et al., 1994). The first is systematic error or bias that would occur on average for subjects measured. The second is subject error, which is additional error that varies from subject to subject. The subject error is also called imprecision or the measure of the variation of measurement error in the population. Precision can be assessed by a construct known as the validity coefficient. It ranges from 0 to 1 with the value one indicating that the observed measurement is a perfectly precise indicator of the true measurement (Armstrong et al., 1994). A validity study would be defined here as one in which a sample of individuals is measured twice: once using the biomarker test of interest and once using a perfect measure of the true biomarker (White, 1997). However, for most biomarkers such perfect measures of the true biomarker do not exist, and, in practice, validation of a method must rely on comparison to other (similarly unvalidated) methods. Then the indicator of biomarker measurement error from the validity study can be applied to what is known about the association under study in the parent study to estimate the effects of biomarker error on the association of interest (White, 1997). While the impact of measurement error on exposure-disease associations has been studied extensively, the impact on estimates of interaction of two or more risk factors has been studied less thoroughly (Greenland, 1993). Assessment of interaction of multiple exposures, gene-environment or gene-gene is an important issue in environmental epidemiology and all the more important with biomarkers depicting mechanistic events.

There are numerous sources of measurement error in biomarkers; some of these are shown in Table 5.

Table 5. Examples of sources of error in biomarker measurement in epidemiological studies[a]

Errors in the laboratory method as a measure of the exposure of interest

- Method may not measure all sources of the biological true exposure of interest

- Method may measure other exposures that are not the true exposure of interest

- Methods may be influenced by subject characteristics (other than the true exposure) that the researcher cannot manipulate, e.g., by the disease under study or by other diseases

Errors or omissions in the protocol

- Failure to specify the protocol in sufficient detail regarding timing and method of specimen collection, specimen handling, storage and laboratory analytical procedures

- Failure to include standardization of the instrument periodically throughout the data collection

Errors due to variation in execution of the protocol

- Variations in method of specimen collection
- Variations in specimen handling or preparation
- Variations in length of specimen storage
- Variations in specimen analysis between batches (different batches of chemicals, different calibration of instrument)
- Variation in technique between laboratory technicians
- Random error within batch

Adapted from: White (1997)

[a] In addition to measurement error, the uncertainty of the results is affected by biological variability within subjects, i.e., short-term variability (hour to hour, day to day) in biological characteristics due to, for example, diurnal variation, time since last meal, posture (sitting vs lying down); medium-term variability (month to month) due to, for example, seasonal changes in diet; and long-term change (year to year) due to, for example, purposeful dietary changes over time.

Ultimately, validation requires the use of epidemiological study designs to assess at least one of three types of relationships: exposure-dose; biological effects-disease; and susceptibility influencing an exposure-disease relationship. Studies that contribute to these types of validation and bridge the gap between laboratory experimentation and population-based epidemiology have been referred to as "transitional" studies (Hulka, 1991; Schulte et al., 1993; Rothman et al., 1995). They may be designed to evaluate exposures, health effects or susceptibility, and some may have the characteristics of pilot or developmental studies (Hulka & Margolin, 1992).

In this section the kinds of information and approaches to validate specific types of biomarkers are discussed. Characteristics of valid biomarkers are outlined in Table 6.

Table 6. Characteristics of valid biomarkers

Biomarker type	Characteristic of validity
Exposure	Consistently linked with exposure at relevant levels of exposure with confounding and background exposures assessed[a]
Effect	Consistently linked with increased risk with confounding and effect modifying factors assessed
Susceptibility	Can distinguish subgroups at risk given specific exposure

[a] Biomarkers of exposure may also be validated by establishing a constant link to an adverse health effect or to the concentration of the chemical in the target organ.

4. VALIDATION OF SPECIFIC TYPES OF BIOMARKERS

4.1 Exposure biomarkers

The validation of biomarkers of exposure requires equal attention to assessing both the exposure and the biomarkers so that a fair comparison between them can occur. However, the relationship between the biomarkers and the exposure will vary due to host factors as the biomarkers become further away from the exposure, depending on the number of steps in the absorption, metabolism and clearance pathways between uptake and the specific biomarker (Schulte & Waters, 1999). This applies to any form of exposure. It is due to intervening host factors that vary between individuals such as breathing rate and capacity, activation, detoxification, elimination, DNA repair, etc. Thus a high correlation between exposure and the marker may not always be observed and an exposure-response relationship may vary between people. It is therefore important to identify and adjust for factors that can influence an exposure-response relationship. For example, to validate hydroxy-ethyl haemoglobin adducts as exposure biomarkers for ethylene oxide at low dose, investigators adjusted for age, smoking, and education in a linear regression model (Schulte et al., 1992). Additionally it may be useful to consider effect modifying factors, such as metabolic polymorphisms (Bois et al., 1995).

There are some exceptions to the validation strategy that focuses on the demonstration of a correspondence between a biomarker of exposure and external exposure. Alternative ways to validate biomarkers include the assessment of their relationship with the concentration in the critical organ (e.g., concentration of cadmium in the kidney vs. in blood or urine) or with critical effects (neurotoxic effects of lead vs. blood lead concentration). Indeed, a good biomarker of exposure should be useful to predict adverse effects, rather than exposure levels. This may be especially the case when accurate and valid measurements of the "true" exposure are difficult or impossible to obtain (use of protective devices, multiple pathways of uptake, etc.).

It is possible to apply qualitative tests to determine whether external exposure or an exposure biomarker would be a better predictor for disease (Steenland et al., 1993). One test involves determining if the biomarker is more highly correlated or associated with the disease than external exposure. A second test is whether, given the same level of exposure, those with higher levels of the biomarkers are more likely to develop the disease.

When absorption mainly occurs through the dermal route or when individual protective devices are used, biomarkers of exposure can provide reliable measurements of internal dose, which are useful to assess dose-response relationships. Fig. 3 shows a logistic regression model based on published data (Calleman et al., 1994) showing that 97.5% of subjects with clinical signs of peripheral neuropathy are correctly classified on the basis of acrylamide adducts to N-terminal valine (AAVAL) in haemoglobin (Hb). On the basis of the parameters of the logistic regression, the calculated benchmark dose corresponds to 0.8 nmol AAVAL/g Hb.

In evaluating the role of metabolic polymorphisms, the presence of a range of doses in which the modifying effect of metabolic enzymes could be seen, is a major issue. A pertinent example comes from a study on the urinary excretion of 1-hydroxypyrene in traffic police officers (Merlo et al., 1998). In the study group, subjects carrying the CYP1A1+ polymorphism had higher levels of hydroxypyrene in the urine, but only at low exposures to PAHs.

4.2 Effect biomarkers

Biomarkers of intermediate effects, i.e., between exposure and disease, can be validated in case-control studies and cohort studies (Rothman et al., 1995; Howe, 1998; Hagmar et al., 1998; Muñoz & Gange, 1998). Once validated, these markers can serve as surrogates for disease, albeit with some probability functions since generally not all people with a given biomarker will develop the disease, but the groups with the high levels generally will be at greatest risk. A good example comes from a recent prospective study on the association between cytogenetic biomarkers and cancer risk (Hagmar et al., 1998). This study, which followed five European cohorts has shown that subjects in the group with the highest frequency of chromosomal aberrations experienced an overall cancer risk more

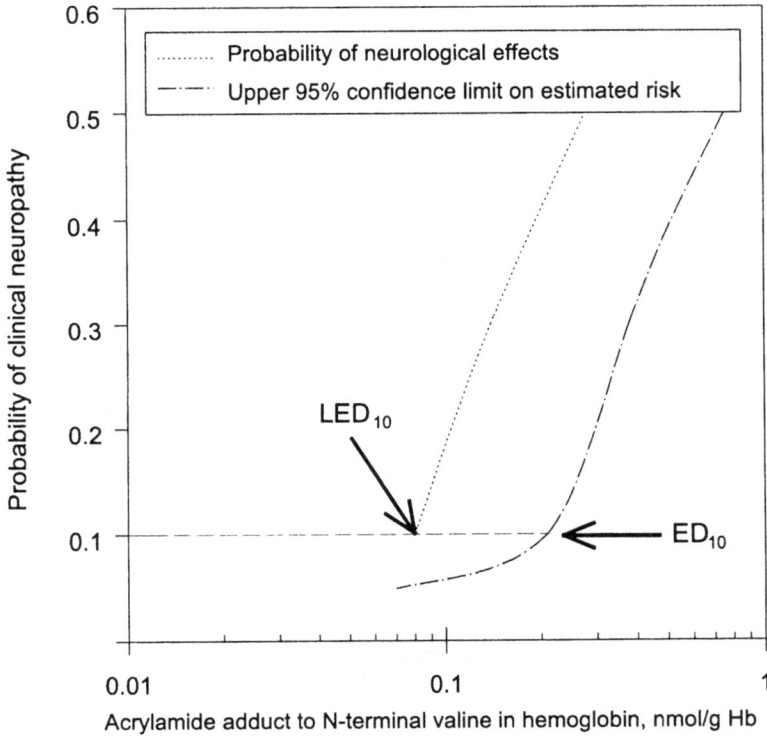

Fig. 3. Acrylamide-induced neurological damage and acrylamide adducts to N-terminal valine in haemoglobin (from data by Calleman et al., 1994)

than double with respect to the lowest frequency group. In the same study, no association was observed between sister chromatid exchange (SCE) frequency and cancer risk, whereas inconclusive results were found for the micronucleus assay. More recently, a nested case-control study found that the association between chromosomal aberrations and cancer appeared to be independent of host factors like age and sex, and could not be explained by exposure to identified human carcinogens (Bonassi et al., 2000).

The lack of validation of most biomarkers of intermediate effect is probably the most critical impediment to the broad use of biomarkers in risk assessment. Validated biomarkers of effect can be used as disease surrogates and thus will be countable end-points that can fill voids left by the inability to count low frequency adverse morbidity or mortality events (Hattis & Silver, 1993; Goldstein, 1996; McMichael & Hall, 1997). Earlier results may be obtained from epidemiological studies if use of a biomarker increases the statistical power of the study (McMichael & Hall, 1997). The prospective epidemiological study is the gold standard for validation effect biomarkers. The timing and frequency of specimen collection in prospective studies are important and can influence the validation. This type of study provides estimates of the risk of disease of individuals with and without a particular biomarker. These studies are time consuming and costly. To reduce the time and cost variables, efforts are underway to follow prospectively large cohorts and bank biological specimens (Willett, 1998). These will allow for "compressed" evaluations of a biomarker and disease risk at the same time.

A measure of the degree of validation of an intermediate marker of effect is the extent to which the exposure is mediated through a marker, i.e. whether the marker is actually strongly predictive of the clinical disease, and the disease never or rarely occurs without this antecedent marker. This may be assessed by calculating the attributable proportion which has also been referred to in the literature as "population attributable risk" or "etiologic fraction" (Benichou, 1991; Trock, 1995). The attributable proportion associated with a particular biomarker is an estimate of the proportion of diseased cases that must progress through the biomarker, i.e., the cases that would not occur if the event(s) resulting in the biomarker could be prevented (Schatzkin et al., 1990; Trock, 1995). The attributable proportion (AP) includes consideration of the sensitivity (S) of the assay (i.e., the ratio of subjects positive in the assay, who developed the disease, to the whole number of diseased subjects), and the relative risk (RR) of disease for the subjects positive in the assay. It is defined as: $AP = S(1-(1/RR))$. The sensitivity is the factor with the greatest impact in the attributable proportion (Schatzkin et al., 1990). The attributable proportion takes into account both the strength of an association between a marker and disease and also the prevalence of the marker. Thus, for example, using data from the European studies on chromosomal aberrations and cancer (sensitivity 46/91 = 0.50, RR 1.53/ 0.79 = 1.93), the proportion of cancer cases attributable to the chromosomal damage was calculated to be 24% for the Nordic cohort (Bonassi, 1999). Trock (1995) has described the next step.

Once it has been established that a significant proportion of tumours can be attributed to a particular marker, epidemiological principles concerned with 'intervening variables' can be used to examine the extent to which the marker truly represents an event intervening between exposure and cancer. In evaluating the relationship between an exposure and a disease outcome, one typically does not use statistical adjustment methods to adjust for a variable that is an intermediate step between exposure and outcome (Weinberg, 1993). Such an adjustment would sharply reduce or even eliminate the apparent effect of the exposure since the marker's association with disease is a direct result of its association with exposure (assuming that the marker represents the relevant time period of exposure with respect to onset of disease) (Trock, 1995). One can take advantage of this property to assess the role of a marker as an intervening variable. If one compares the crude (i.e., unadjusted) RR for exposure to the RR for the exposure effect adjusted for the biological marker, the extent to which adjustment for the marker has reduced the apparent exposure effect indicates the degree to which the marker is linked to the exposure-disease relationship (Trock, 1995). If the effect of exposure occurs primarily through a pathway involving the marker, then the marker-adjusted exposure effect will essentially be eliminated, i.e., the adjusted RR will be close to 1.0 (Schatzkin et al., 1990). In some cases a marker will be useful even though it is not on the causal pathway. Those are cases where it correlates to something on the causal pathway (e.g., protein adducts, Ehrenberg et al., 1996) and ultimately to disease risk.

Another measure of validation of a biological marker of effect is the positive predictive value. Predictive value for a marker of disease is the proportion of people studied with a particular disease among all the people who have the marker. Predictive value is not only a

25

property of the marker assay, it is determined by the sensitivity and specificity of the assay and the prevalence of the disease. Thus, for example, a marker that is 90% sensitive and 90% specific will still only have a positive predictive value of 50% when the prevalence of the underlying disease is 10%. Parallel considerations should be extended to the use of the negative predictive value whenever the hypothesis of association is rejected. Field studies that do not incorporate prevalence considerations in planning are likely not to be able to detect an association between a marker of effect and disease, even if one exists (Schulte & Perera, 1993).

Positive predictive value and attributable proportion reflect very different things (Ottman, 1995; Khoury & Wagener, 1995). Positive predictive value, the risk of disease among persons with a specific marker is important from the point of view of the individual. Attributable proportion, on the other hand is the proportion of diseased cases that must progress through the biomarkers and thus could be prevented if that process could be interrupted. This is important from an environmental or public health point of view.

In addition to positive predictive value and attributable proportion, two other concepts are useful in the interpretation of biomarker data: negative predictive value (see above) and sentinel biomarkers. The concept of a sentinel biomarker involves a biomarker that, regardless of predictive value or attributable proportion, may have properties (increased frequency of increased concentration or of occurrence) that might be indicative of exposure to an environmental hazard or onset of a biological effect (see Appendix 2).

4.3 Susceptibility biomarkers

Considerable variability exists in the response of humans to toxic substances. A very large number of genetic conditions potentially enhancing one's susceptibility to chemicals have been identified (Appendix 3). However, only in a few cases such as the glucose-6-phosphate dehydrogenase deficiency, has a causal relationship been demonstrated, deficient individuals being more susceptible to toxic environmental oxidants (Stokinger & Mountain, 1963).

Polymorphisms may be markers of susceptibility and a long list of genes and their variants (polymorphisms) has been and will be established. Many of these genes are quite general in function, but changes in this function due to a polymorphism may influence the susceptibility of developing disease. In addition to polymorphisms in xenobiotica metabolism enzymes, polymorphisms in genes that influence or control cell differentiation, apoptosis, cell cycle kinetics, signal transduction and DNA repair may influence the health outcome when exposed to an environmental toxicant.

Perhaps the greatest potential contributions of biomarkers to risk assessment and risk management will be the inclusion of inherited susceptibility biomarkers. However, while there has been extensive use of susceptibility biomarkers in the development of pharma-ceuticals (Evans & Relling, 1999), the potential contribution to risk assessment in occupational or environmental chemical exposure has rarely been realized. Susceptibility biomarkers may reflect variation in exposure, kinetics and effects, and are therefore important to consider in risk assessments (Bois et al., 1995; Dickey et al., 1997). These biomarkers have both promises and perils for individual and population risk estimation. The promise is for a more refined assessment of risk through the identification of gene-gene and gene-environment interactions and also for the focusing of prevention and control programmes on high-risk individuals. The perils include ethical and social issues including stigmatization, discrimination and the misconception that removing a susceptible person from the exposure scenario without reducing exposure opportunities will reduce risk effectively, when it may not, on a comparative basis (Vineis & Schulte, 1995). There are also issues in using susceptibility markers as effect modifiers in epidemiological studies. These include: whether there is a correspondence between a genotype and phenotype; whether there is a mechanistic reason to consider the marker; and whether the prevalence of the allele is frequent enough to assess in the population in a practical way.

Before applying susceptibility biomarkers in epidemiological studies, there are several important issues to consider. The use of genotype rather than phenotype may lead to misclassification as the actual enzyme activity is influenced by a number of exogenous and endogenous factors, thus diluting the genetic component. Further-more, susceptibility factors, be they enzymes, receptors or other target molecules, are highly compound-specific, so that even closely related substances may not be substrate to the same polymorphic enzymes. Combination of and interactions between various alternative pathways governed by polymorphic and environmentally regulated enzymes are important as metabolism generally involves several stages. The effect of the genetic polymorphism may also depend on the outcome measure. For example, in a case report, workers exposed to methyl bromide showed different neurotoxic responses, people expressing the gene GSTT1 being more severely affected. In contrast, the least sensitive to the neurotoxic effect had the highest level of alkylation damage (Garnier et al., 1996).

One of the most widely known examples of how susceptibility biomarkers combined with an exposure measure can inform risk assessment is the metabolic polymorphism for *N*-acetyltransferase in the case of bladder cancer. This enzyme is involved in the detoxification of arylamines, and individuals classified as slow acetylators have an increased risk of bladder cancer when exposed to e.g., beta-naphthylamine (Cartwright et al., 1982). However, in a Chinese study the NAT genotype did not influence the risk of bladder cancer, but in this situation the exposure was to benzidine. Monoacetylation mediated by NAT is an activation, rather than a deactivation pathway of benzidine (Rothman et al., 1996). This stresses the importance of knowledge of the biotransformation of the compound under investigation prior to testing the role of susceptibility markers.

The influence of genetic factors on exposure-disease associations can be large. Calabrese (1997) demonstrated that genetically determined biochemical differences between people for a range of phenotypes could exceed 10-fold. In a Monte-Carlo simulation study, Bois et al. (1995) illustrated, by pharmacokinetic modelling of DNA adducts in the bladder of people exposed to oral bolus doses of 4-aminobiphenyl, that the adduct levels of the most susceptible individuals are 10 000 times higher than for the least susceptible and that the 5th and 95th percentiles differ by a factor of 160. Input parameters for the model were derived from the literature on human *in vitro* studies, or from dogs in the absence of human data. Therefore, accounting for genetic variability may have important implications for risk assessment (Dickey et al., 1997).

The assessment of gene-environment interaction is important in the validation of biomarkers of susceptibility (and exposure). Gene-environment interaction is defined as "a different effect of an environmental exposure on disease risk in persons with different genotypes," or alternatively, "a different effect of genotype on disease risk in persons with different environmental exposures" (Ottman, 1996). Ottman (1996) has described five biologically plausible models of gene-environment interaction, each of which leads to a different set of predictions about disease risk in individuals classified by the presence or absence of a high-risk genotype or environmental exposure. If a biomarker of susceptibility is to be validated for disease, its relationship to both disease and exposure needs to be determined. Valid evaluation of gene-environment interaction requires the accurate measurement of both genetic and environmental factors (Rothman et al., 1999). Modest exposure assessment errors may produce a biased estimate of the interaction parameter that results in a substantial increase in sample size requirements (Garcia-Closas et al., 1998).

5. CROSS-SPECIES COMPARABILITY

The optimal use of biomarkers in environmental health risk assessments will most likely occur if human studies are linked to studies of laboratory animals and cell lines (Shugart et al., 1992; Anderson S et al., 1994). An additional extension of the use of biomarkers is studying appropriate species of wildlife (Barrett et al., 1997). Biomarkers can serve as a common element in studies of these different groups or materials. Thus a biomarker identified in an exposed laboratory animal or cell line might also be seen in wild or laboratory animals or humans with similar exposures.

A parallelogram type approach (Sobels, 1993; Sutter, 1995) can be used to assess the relationship between markers and risks in those groups (Fig. 4). The parallelogram approach is derived from the work in the 1970s of Sobels (1993) to extrapolate genetic damage from animals to humans. Genetic damage which cannot be measured directly, such as in human germ cells, can be estimated by measuring the same kind of damage in both germ cells and somatic cells of the mouse. With data on the induction of mutations or chromosomal aberrations in both germ cells and somatic cells of the mouse, it is possible to estimate germ cell mutation frequencies in humans on the basis of what can be measured by monitoring genetic damage in human somatic cells (Sobels, 1993). Sutter (1995) has modified this approach to include *in vitro-in vivo* extrapolation. In the parallelogram experimental approach to knowledge of mechanism, *in vitro* data are used to test the hypothesis that a specific mechanism of action exists in rodents and humans (Sutter, 1995).

An example where the original parallellogram approach has been used is for the genotoxic compound 1,3-butadiene (Pacchierotti et al., 1998). The purported estimates of heritable damage in man were based on data for heritable translocations in germ cells, and bone marrow micronuclei induced in mice and chromosomal aberrations on lymphocytes of exposed males. The rate of heritable translocation induction per ppm/h of butadiene exposure was estimated to be approximately 0.8 per million live born compared to spontaneous incidence of balanced translocations in humans of approximately 800 per million born. Other compounds have also

Mouse cancer

detected tumour
incidence and
adducts

Human cancer

detected cancer
incidence and
adducts

Mouse somatic cells

measured mutations
and adducts

Human somatic cells

measured mutations
and adducts

Mouse germ cells

measured mutations
and adducts

Human germ cells

estimated mutations

Comparisons

Estimates

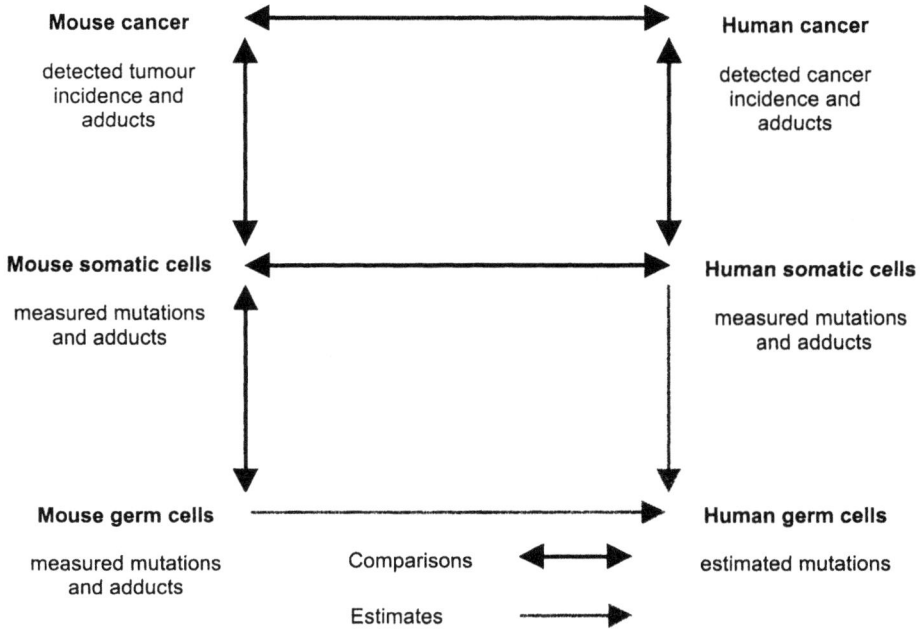

Fig. 4. The enhanced parallelogram concept.

been characterized using the parallellogram approach, such as ethylene oxide, cyclophosphamide and acrylamide (Waters & Nolan, 1995).

Clearly, most identified human carcinogens are genotoxic, thus helping to build the case for human germ cell mutagenicity. However, there are 13 putative germ cell non-mutagens adequately tested for carcinogenicity, 11 of which are genotoxic carcinogens and these include vinyl chloride and propylene oxide (Waters et al., 1999).

With regard to endocrine disruptors, those operating through the spindle receptor or the genotoxic estrogens such as fosfesterol could usefully be examined using the parallellogram approach to determine the germ cell risks (DeRosa et al., 1998; Olea et al., 1998).

Biomarkers can be used to reduce high- to low-dose and species extrapolation-related uncertainties by providing information on common mechanisms and the development of mechanistically based mathematical models (Sexton et al., 1995). The incorporation of biomarkers of exposure and susceptibility in physiologically based pharmacokinetic models has allowed for interspecies comparisons and enabled the simulation of different enzyme activities among individuals (Fennell et al., 1996). Biomarkers also may serve as an alternative to the use of PBPK models for determining dose (Rhomberg, 1995). They are particularly useful when they are more easily or accurately measured than the actual exposure. Since some biomarkers have extraordinary sensitivity, they may significantly extend the range of empirical characterization of dose and response in cases where they may be detected and measured at dose levels below those at which other effects are directly observable (Rhomberg, 1995; Ehrenberg et al., 1996). For example, adduct measurements of some alkylating agents may be used to indicate disease risks at levels too low to be detected by epidemiological means (Ehrenberg et al., 1996).

6. NEW PERSPECTIVES

Validation and successful use of biomarkers require a high degree of analytical accuracy and knowledge of what they mean in terms of health and disease.

Recent developments in information technology, molecular biology and instrumentation have provided new tools for use in environmental health research and biologically based risk assessment.

The specificity and sensitivity of many biomarkers will be improved by the introduction of new analytical methodologies, e.g., speciation of metal ions by inductively coupled plasma mass spectrometry (ICP-MS) and mass-spectrometric techniques to detect metabolites and adducts.Chromosomal aberration appears to be one of the most promising biomarkers for its association with cancer risk. Application of new detection methods, i.e., primed *in situ* labelling (PRINS) and fluorescence *in situ* hybridization (FISH) will extend the observation from the chromosome level to specific genes relevant for the disease process. Imaging technologies such as magnetic resonance or positron emission tomography (PET) and single photon emission computerized tomography (SPECT) are particularly interesting for studies in human populations as these methods are non-invasive and can measure change at the molecular scale.

Environmental genome projects will identify new single nucleotide polymorphisms (SNP) in genes involved in the disease process, information that can be used to identify new susceptibility factors, e.g., control of cell cycle and apoptosis. A spin-off of the project has been the development of new instrumentation making large-scale epidemiological studies using genetic markers of susceptibility more feasible, including the study of gene-gene interaction. Detection of polymorphisms in almost any protein will result in large amounts of data on their biological and health effects. The data will help detect new susceptible populations. The explosion of polymorphism data requires extension of bioinformatic approaches towards epidemiological databases.

Development of new animal and *in vitro* models, e.g., expressing specific allelic variant, will enable us to study the role of specific enzymes as risk factors in disease development and on the level of biomarkers. Knowledge of metabolism, product formation and general mechanisms of action are required for the development of biomarkers for environmental agents.

High output technologies, such as DNA microarray, can be used to study gene function and expression. The technologies will allow for evaluation of the temporal and spatial pattern of gene expression under various exposure conditions. Furthermore, they will give additional information on response in different species, thus validating animal models for the study of human disease. The technology could lead to better characterization and understanding of the disease processes, and the development of new relevant biomarkers.

Introduction of new and validated methods based upon new technologies to study biomarkers of exposure, effect and suscepti-bility at the different levels of the risk management process will be of great assistance to risk assessors.

Advances in biologically based pharmacokinetic modelling and simulations and new approaches for addressing uncertainty will be a major way for incorporating biomarkers into rich assessments. PBPK models are becoming increasingly complex and uncertainty analyses for models of such complexity are difficult to apply by using analytical calculation. Stochastic models such as Monte Carlo simulations and Markovchain Monte Carlo models are useful to address these difficulties. However Monte Carlo simulations have limitations, such as those involving prior distributions for model parameters. Bayesian analysis is a useful approach for addressing these distributions (Bois, 1999).

7. SUMMARY

The Task Group, building on previous categorizations and evaluations of biomarkers for research, considered them for risk assessment. Risk assessment was defined as the set of steps between research and risk management. It provides society with estimates of risk when uncertainty exists about the safety of prevailing or future levels of exposure to environmental and occupational toxicants.

A framework for selecting and validating biomarkers for risk assessment was developed by the Task Group. Examples were cited of how the three types of biomarkers, of exposure, of effect and of susceptibility, could be validated for research and be used in risk assessments. Valid biomarkers can lead to biologically based risk assessments.

There have been few instances where validated biomarkers have been used in quantitative risk assessments. Future work should include scientific, technical, organizational and administrative efforts to coordinate efforts to set an agenda for research on biomarkers that will contribute to conducting important risk assessments. This will require long-term commitments for collaboration and the conduct of prospective studies to link biomarkers to disease risks.

8. CONCLUSIONS

Validated biomarkers are useful in reducing uncertainty in risk assessments. However, biomarkers should be viewed as another set of tools available for researchers and risk assessors, not as a replacement for traditional approaches.

Validation of biomarkers for research and risk assessment requires both laboratory and epidemiological studies.

Successful use of biomarker data implies an understanding of mechanism. The incorporation of mechanistic data in risk assessment is certainly important, but risk assessments and regulations should not wait for the development of mechanistic data nor should uncertainty about mechanism be used to block public health action.

The contribution of biomarkers of susceptibility has great potential but has yet to be realized on a large scale in quantitative risk assessment.

There is a need for a long-term commitment to the assessment of the validity of biomarkers for risk assessment, environmental health research and public health practice.

9. RECOMMENDATIONS

In making the following recommendations, the Task Group recognized the role given to the IPCS to facilitate and increase coordination of international activities in order to promote the further work needed to define human health effects associated with exposure to chemicals and to provide the basis for priority-setting actions in order to protect health.

9.1 General recommendations

The following recommendations were formulated:

- to formulate ethical guidelines to promote biomarker research while guaranteeing individual privacy and integrity;

- to incorporate mechanistic information, utilizing bio-markers, in risk assessment;

- to develop an international set of principles for the collection, archiving and use of biological specimens and acquired data, notably those from humans;

- to establish guidelines for biomarker studies in humans and to critically evaluate whether and under what conditions such experimentation is warranted.

9.2 Recommendations for future research

9.2.1 Prevalidation stage

The Task Group noted that results from validated biomarkers should be used in public health decisions, and recommended:

- to develop more incisive biomarkers to fill in the gaps in the continuum of events from environmental exposure to clinical disease expression, taking advantage of new, high-throughput technologies;

- to study the genetic basis for different susceptibilities toward environmental exposures and how this exposure influences the phenotype;

- to develop and use bioinformatics and advanced statistical methods to fully utilize existing and newly generated data.

9.2.2 Validation stage

The Task Group further recommended:

- to better characterize biomarkers with respect to their sensitivity and specificity, and to validate their predictability for adverse health effects;

- to validate biomarkers in high-quality analytical epidemiological studies; prospective studies are likely to give the most definitive validation;

- to promote validation of biomarkers, researchers should store information and specimens for pooled or subsequent analysis;

- to model human and experimental data to link biomarkers with the expression of disease.

9.3 Application

The Task Group recommended:

- to study the population distribution and role in disease development of risk factors using validated susceptibility biomarkers;

- to use validated biomarkers to study the effects of public health prevention initiatives among populations exposed to toxicants.

REFERENCES

Albertini RJ, Anderson D, Douglas GR, Hagmar L, Hemminki K, Merlo F, Natarajan AT, Norppa H, Shuker DE, Tice R, Waters MD, & Aitio A (2000) IPCS guidelines for the monitoring of genotoxic effects of carcinogens in humans. Mutat Res, **463**: 111-172.

Andersen ME, Clewell HJ, Gargas ML, MacNaughton MG, Reitz RH, Nolan RJ, & McKenna MJ (1991) Physiologically based pharmacokinetic modeling with dichloromethane, its metabolite, carbon monoxide, and blood carboxyhemoglobin in rats and humans. Toxicol Appl Pharmacol, **108**: 14-27.

Anderson D, Sorsa M, & Waters MD (1994) The parallelogram approach in studies of genotoxic effects. Mutat Res, **313**: 101-115.

Anderson S, Sadinski W, Shugart L, Brussard P, Depledge M, Ford T, Hose J, Stegemen J, Suk W, Wirgn I, & Wogan G (1994) Genetic and molecular ecotoxicology: a research framework. Environ Health Perspect, **102** (Suppl 12): 3-8.

Armstrong BK, White E, & Saracci R (1994) Principles of exposure measurement in epidemiology. New York, Oxford University Press, pp 49–114.

Bailar JC III & Bailer AJ (1999) Common themes at the workshop on uncertainty in the risk assessment of environmental and occupational hazards. In: Bailer AJ, Maltoni C, Bailar JC III, Belpoggi F, Braz JV, & Soffritti M ed. Uncertainty in the risk assessment of environmental and occupational hazards. Ann NY Acad Sci, **895**: 373-376.

Bailer AJ & Dankovic DA (1997) An introduction to the use of physiologically based pharmacokinetic models in risk assessment. Stat Methods Med Res, 6: 341-358.

Barrett J, Vainio H, Peakall D, & Goldstein BD ed. (1997) 12th Meeting of the scientific group on methodologies for the safety evaluation of chemicals: susceptibility to environmental hazards. Environ Health Perspect, **105**: 699-737.

Becking GC (1995) Use of mechanistic information in risk assessment for toxic chemicals. Tox Letters, **77**: 15-24.

Benichou J (1991) Methods of adjustment for estimating the attributable risk in case-control studies: a review. Statistics in Medicine, **IV**: 1753-1773.

Bernard AM (1995) Biokinetics and stability aspects of biomarkers: recommendation for applications in population studies. Toxicology, **101**: 65-71.

Boffetta P (1995) Sources of bias, effect of confounding in the application of biomarkers to epidemiological studies. Toxicol Lett, **77**: 235-238.

Bois FY (1999) Analysis of PBPK models for risk characterization. Ann. N Y Acad Sci, **895**: 317-337.

Bois FY, Krowech G, & Zeise L (1995) Modeling human interindividual variability in metabolism and risk: The example of 4-aminobiphenyl. Risk Analysis, **15**: 205-213.

Bonassi S (1999) Combining environmental exposure and genetic effect measurements in health outcome assessment. Mutat Res, **428**: 177-185.

Bonassi S, Ceppi M, Fontana V, & Merlo F (1994) Multiple regression analysis of cytogenetic human data. Mutat Res, **313**: 69-80.

Bonassi S, Hagmar L, Strömberg U, Montagud AH, Tinnerberg H, Forni A, Heikkilä P, Wanders S, Wilhardt P, Hansteen I-L, Knudsen LE, & Norppa H (2000) Chromosomal aberrations in lymphocytes predict human cancer independently of exposure to carcinogens. Cancer Res, **60**: 1619-1625.

Calabrese EJ (1997) Role of genetic factors in environmentally induced toxic response: historical consideration, present status and future directions. University of Massachusetts, Amherst, MA.

Calleman CJ, Wu Y, He F, Tian G, Bergmark E, Zhang S, Deng H, Wang Y, Crofton KM, Fennell T, & Costa LG (1994) Relationship between biomarkers of exposure and neurological effects in a group of workers exposed to biomarkers of exposure and neurological effects in a group of workers exposed to acrylamide. Toxicol Appl Pharmaco, **126**: 361-371.

Cartwright RA, Glashan RW, Rogers HJ, Ahmad RA, Barham-Hall D, Higgins E, & Kahn MA (1982) Role of N-acetyltransferase phenotype in bladder carcinogenesis: a pharmacogenetic epidemiological approach to bladder cancer. Lancet, **11**: 842-846.

Dary CC, Quackenboss JJ, Nauman CH, & Hern SC (1996) Relationship of biomarkers of exposure to risk assessment and risk management. In: Blancato JN, Brown RN, Dary CC, Saleh MA ed. Biomarkers for agrochemicals and toxic substances. ACS Symposium Series 643, ACS, Washington, DC, pp 2-23.

DeRosa C, Richter P, Pohl H, & Jones DE (1998) Environmental exposures that affect the endocrine system: public health implications. J Toxicol Environ Health B Crit Rev, **1**: 3-26.

D'Errico A, Taioli E, Chen X, & Vineis P (1996) Genetic metabolic polymorphisms and the risk of cancer: a review of the literature. Biomarkers, **1**: 149-173.

Dickey C, Santella RM, Hattis D, Tabg D, Hsu Y, Cooper T, Young T-L, & Perera FP (1997) Variability in PAH-DNA adduct measurements in peripheral mononuclear cells: implications for quantitative cancer risk assessment. Risk Anal, **17**: 649-654.

Dor F, Dab W, Empereur-Bissonnet P, & Zmirou D (1999) Validity of biomarkers in environmental health studies: The case of PAHs and Benzene. Critical Rev Tox, **29**: 129-168.

Ehrenberg L, Granath F, & Törnqvist (1996) Macromolecular adducts as biomarkers of exposure to environmental mutagens in human populations. Environ Health Perspect, **104** (Suppl 3): 423-428.

El-Masri HA, Bell DA, & Portier CJ (1999) Effects of glutathione transferase theta polymorphism on the risk estimates of dichloromethane to humans. Toxicol Appl Pharmacol, **158**: 221-230.

Emancipator K (1999) Laboratory diagnosis and monitoring of diabetes mellitus. Am J Clin Pathol, **112**: 665-674.

Evans WE & Relling MV (1999) Pharmacogenomics: translating functional genomics into rational therapeutics. Science, **286**: 487-491.

Fennell TR, MacNeela JP, Thompson CL, & Bell DA (1996) Hemoglobin adducts from acrylonitrile and ethylene oxide in cigarette smokers: effects of glutathione transferase TI and MI genotypes. Toxicologist, 30, 282 [Abstract No. 1443].

Garcia-Closas M, Rothman N, Stewart WF, & Lubin JH (1998) Impact of misclassification in studies of gene-environmental interactions. Proc Am Assoc Cancer Res, **39**: 181.

Garnier R, Rambourg-Schepens MO, Muller A, & Hallier E (1996) Glutathione transferase activity and formation of macromolecular adducts in two cases of acute methyl bromide poisoning. Occup Environ Med, **53**: 211-215.

Goldberg M & Hémon, D (1993) Occupational epidemiology and assessment of exposure. Int J Epidemiol, **22** (Suppl 2): s5-s9.

Goldstein BD (1996) Biological markers and risk assessment. Drug Metab Rev, **28**: 225-233.

Greenblatt M, Bennett W, Hollstein M, & Harris C (1994) Mutations in the p53 tumor suppressor gene: clues to cancer etiology and molecular pathogenesis. Cancer Res, **54**: 4855-4878.

Greenland S (1993) Basic problems in interaction assessment. Environ Health Perspect, **101**(Suppl 4): 59-66.

Hagmar L, Bonassi S, Strömberg U, Brøgger A, Knudsen LE, Norppa H, Reuterwall C, and the European Study Group on Cytogenetic Biomarkers and Health (1998) Chromosomal aberrations in lymphocytes predict human cancer: a report from the European Study Group on Cytogenic Biomarkers and Health (ESCH). Cancer Res, **58**: 4117-4126.

Hatch M & Thomas D (1993) Measurement issues in environmental epidemiology. Environ Health Perspect, **101**(Suppl 4): 49-57.

Hattis D & Silver K (1993) Use of biomarkers in risk assessment. In: Schulte PA & Perera FP ed. Molecular epidemiology: principles and practices, San Diego, Academic Press, pp 251-273.

Hernberg S & Aitio A (1987) Validation of biological monitoring tests. In: Foa V, Emmett EA, Maroni M, & Columbi A ed. Occupational and environmental chemical hazards: cellular and biochemical indices for monitoring toxicity. Chichester, England, Ellis Horwood, pp 41-49.

41

Howe GR (1998) Practical uses of biomarkers in population studies. In: Mendelsohn ML, Mohr LC, & Peeters JP ed. Biomarkers: medical and workplace applications, pp 41-49.

Hulka BS (1991) Epidemiological studies using biological markers: issues for epidemiologists. Cancer Epidemiol Biomarkers Prev, 1: 13-19.

Hulka BS & Margolin BH (1992) Methodologic issues in epidemiological studies using biomarkers. Am J Epidemiol, **135**: 200-204.

IARC (1997) Application of biomarkers in cancer epidemiology. Lyon, International Agency for Research on Cancer, IARC Scientific Publications No. 142, p1.

IPCS (1992) Environmental Health Criteria 134: Cadmium. Geneva, World Health Organization, International Programme on Chemical Safety, 280 pp.

IPCS (1993) Environmental Health Criteria 155: Biomarkers and risk assessment: Concepts and principles. Geneva, World Health Organization, International Programme on Chemical Safety, 82 pp.

IPCS (1994) Environmental Health Criteria 170: Assessing human health risks of chemicals: Derivation of guidance values for health-based exposure limits. Geneva, World Health Organization, International Programme on Chemical Safety, 73 pp.

IPCS (1999) Environmental Health Criteria 210: Principles for the assessment of risks to human health from exposure to chemicals. Geneva, World Health Organization, International Programme on Chemical Safety, 110 pp.

Joksić G & Spasojević-Tišma V (1998) Chromosome analysis of lymphocytes from radiation workers in the tritium-applying industry. Int Arch Occupa Environ Health, **71**: 213-220.

Khoury MJ & Wagener DK (1995) Epidemiological evaluation of the use of genetics to improve the predictive value of disease risk factors. Am J Genet, **56**: 835-844.

Kriebel D (1994) The dosimetric model in occupational and environmental epidemiology. Occ Hyg, 1: 55-68.

Louis TA (1988) General methods for analyzing repeated measures. Stat Med, **7**: 39-45.

Lucier GW & Thompson CL (1987) Issues in biochemical applications to risk assessment. When can lymphocytes be used as surrogate markers. Environ Health Perspect, **76**: 187-191.

Margetts BM (1991) Basic issues in designing and interpreting epidemiological research. In: Margetts BM & Nelson M ed. Design concepts in nutritional epidemiology. New York, Oxford University Press, pp 13-51.

McClellan RO (1995) Risk assessment and biological mechanisms, lesson learned, future opportunities. Toxicology, **102**: 239-258.

McClellan RO (1999) Human health risk assessment: a historical overview and alternative paths forward. Inhalation Tox, **11**: 477-518.

McMichael AJ & Hall AJ (1997) The use of biological markers as predictive early-outcome measures in epidemiological research. In: Toniolo P, Boffetta P, Shuker DEG, Rothman N, Hulka B, & Pearce N ed. Application of biomarkers in cancer epidemiology. IARC Scientific Publications No. 142, Lyon, International Agency for Research on Cancer, pp 281-289.

Merlo F, Andreassen Å, Weston A, Pab C-F, Haugen A, Valerio F, Reggiardo G, Fontana V, Garte S, Puntoni R, & Abbondandolo A (1998) Urinary excretion of 1-hydroxypyrene as a marker for exposure to urban air levels of polycyclic aromatic hydrocarbons. Cancer Epidemiol Biomarkers Prev, **7**: 147-155.

Morgenstern H & Thomas D (1993) Principles of study design in environmental epidemiology. Environ Health Perspect, **101**(Suppl 4): 23–38.

Muñoz A & Gange SJ (1998) Methodological issues for biomarkers and intermediate outcomes in cohort studies. Epidemiologic Rev, **20**: 29-42.

Muscat JE (1996) Epidemiological reasoning and biological rationale. Biomarkers, **1**: 144-145.

Nelson DI (1997) Risk assessment in the workplace. In: DiNardi S ed. The occupational environment - its evaluation and control. American Industrial Hygiene Association Press, 328-359.

NRC (US National Research Council) (1983). Risk assessment in the Federal Government: Managing the process. Washington, DC, National Academy Press.

NRC (US National Research Council) (1987) Pharmacokinetics in Risk Assessment. Drinking Water and Health, Washington, DC, National Academy Press, p 47.

OSHA (US Occupational Safety & Health Administration) (1998) Methylene chloride; final rule. Federal Register, **63**: 50711-50732.

Olea N, Pazos P, & Exposito J (1998) Inadvertent exposure to xenoestrogens. Eur J Cancer Prev, **7**(Suppl 1): S17-23.

Ottman R (1995) Gene-environment interaction and public health. Am J Hum Genet, **56**: 821-823.

Ottman R (1996) Gene-environment interaction: definition and study designs. Prev Med, **25**: 764-770.

Pacchierotti F, Adler I-D, Anderson D, Brinkworth M, Demopoulos NA, Lähdetie J, Osterman-Golkar S, Peltonen K, Russo A, Tates A, & Waters R (1998) Genetic effects of 1,3-butadiene and associated risk for heritable damage. Mutat Res, **397**: 93-115.

Pearce N, deSanjose S, Boffetta P, Kogevina M, Saracci R, & Savistz D (1995) Limitations of biomarkers of exposure in cancer epidemiology. Epidemiology, **6**: 190-194.

Perera FP (1995) Molecular epidemiology and prevention of cancer. Environ Health Perspect, **103**(Suppl 8): 233-236.

Perera F, Mayer J, Jaretzki A, Hearne S, Brenner D, Young TL, Fishman H, & Grimes M (1989) Comparison of DNA adducts and sister chromatid exchanges in lung cancer cases and controls. Cancer Res, **49**: 4446-4451.

Ponce RA, Bartell SM, Kavanagh TK, Woods JS, Griffith WC, Lee RC, Takaro TK, & Faustman EM (1998) Uncertainty analysis for comparing predictive models of biomarkers: a case study of dietary methyl mercury exposure. Reg Tox Pharm, **28**: 96-105.

Qian GS, Ross RK, Yu MC, Yuan JM, Gao YT, Henderson BE, Wogan GN, & Groopman JD (1994) A follow-up study of urinary aflatoxin exposure and liver cancer risk in Shanghai, People's Republic of China. Cancer Epidemiol Biomarkers Prev, **3**: 3-10.

Remuzzi G & Bertani T (1998) Pathophysiology of progressive renal disease. N Engl J Med, **339**: 1448-1456.

Rhomberg L (1995) Estimation and evaluation of dose. In: Farland W, Olin S, Park C, Rhomberg L, Scheuplein R, Starr T, & Wilson J ed. Low-dose extrapolation of cancer risks: issues and perspectives. Washington, DC, International Life Sciences Institute Press, pp 61-74.

Rothman N, Stewart WF, & Schulte, PA (1995) Incorporating biomarkers into cancer epidemiology: a matrix of biomarker and study design categories. Cancer Epid Biomarkers Prevention, **4**: 301-311.

Rothman N, Bhatnagar VK, Hayes RB, Zenser TV, Kashyap SK, Butler MA, Bell DA, Lakshmi V, Jaeger M, Kashyap R, Hirvonen A, Schulte PA, Dosemeci M, Hsu F, Parikh DJ, Davis BB, & Talaska G (1996) The impact of interindividual variation in NAT2 activity on benzidine urinary metabolites and urothelial DNA adducts in exposed workers. Proc Natl Acad Sci USA, **93**: 5084-5089.

Rothman N, Caporaso NE, Wacholder S, Garcia-Closas M, Lubin JH, Marcus P, Hoover RN, & Fraumeni Jr. JF (1999) Evaluation of interactions between environmental and common genetic polymorphisms: a population-based epidemio-logic perspective. (Abstract). American Association of Cancer Research Annual Meeting, Philadelphia.

Saracci R (1997) Comparing measurements of biomarkers with other measurements of exposure. In: Toniolo P, Boffetta P, Shuker DEG, Rothman N, Hulka B, & Pearce N ed. Application of biomarkers in cancer epidemiology. IARC Scientific Publications No. 142, Lyon, International Agency for Research on Cancer, pp 303-312.

Schatzkin A, Freedman LS, Schiffman MH, & Dawsey SJ (1990) Validation of intermediate endpoints in cancer research. JNCI, **82**: 1746–1752.

Schulte PA (1989) A conceptual framework for the validation and use of biomarkers. Environ Res, **48**: 129-144.

Schulte PA & Mazzuckelli LF (1991) Validation of biological markers for quantitative risk assessment. Environ Health Perspect, **90**: 239-246.

Schulte PA & Perera FP (1993) Validation. In: Schulte PA & Perera FP ed. Molecular epidemiology: principles and practices, San Diego, CA, Academic Press, pp 79-107.

Schulte PA & Talaska G (1995) Validity criteria for use of biological markers of exposure to chemical agents in environmental epidemiology. Toxicol, **101**: 73-88.

Schulte PA & Waters, M (1999) Using molecular epidemiology in assessing exposure for risk assessment. Ann NY Acad Sci, **895**: 101-111.

Schulte PA, Boeniger M, Walker J., Schober SE, Pereira MA, Gulati DK, Wojciechowski JP, Garza A, Froelich R, Strauss G, Halperin WE, Herrick R, & Griffith J (1992) Biological markers in hospital workers exposed to low levels of ethylene oxide. Muta Res, **278**: 237-251.

Schulte PA, Rothman N, & Schottenfield D (1993) Design consideration in molecular epidemiology. In: Schulte PA & Perera FP ed. Molecular epidemiology: principles and practices, San Diego, CA, Academic Press, pp 159-198.

Sexton K, Reiter LW, & Zenick H (1995) Research to strengthen the scientific basis for health risk assessment: a survey for the context and rationale for mechanistically based methods and models. Toxicology, **102**: 3-20.

Shugart LR, McCarthy JF, & Halbrook RS (1992) Biological markers of environmental and ecological contamination: An overview. Risk Analysis, **12**: 353-360.

Sobels FH (1977) Some problems associated with the testing for environmental mutagens and a prospective for studies in comparative mutagenesis. Mutat Res, **46**: 245-260.

Sobels FH (1993) Approaches to assessing genetic risks from exposure to chemicals. Environ Health Perspect, **101**(Suppl 3): 327-332.

Stayner L (1992) Methodologic issues in using epidemiologic studies for quantitative assessment. In: Clewell HJ ed. Proceedings from Conference of Chemical Risk Assessment in the DOD: Science, Policy, and Practice. ACGIH, Cincinnati, OH, pp 43-51.

Steenland K, Tucker J, & Salvan A (1993) Problems in assessing the relative predictive value of internal markers versus external exposure in chronic disease epidemiology. Cancer Epidemiol Biomarkers Prev, **2**: 487–491.

Stewart PA, Blair A, Dosemeci M, & Gomez M (1991) Collection of exposure data for retrospective occupational epidemiologic studies. Appl Occup Environ Hygiene, **6**: 280-289.

Stokinger HE & Mountain JT (1963) Tests for hypersusceptibility to hemolytic chemicals. Arch Environ Health, **6**: 495-502.

Sutter JR (1995) Molecular and cellular approaches to extrapolation for risk assessment. Environ Health Perspect, **103**: 386-389.

Trock BJ (1995) Application of biological markers in cancer environmental epidemiology. Toxicology, **101**: 93-98.

Verberk M (1995) Biomarkers of exposure versus parameters of external exposure; practical applications in estimating health risks. Toxicology, **101**: 107-115.

Vineis P & Schulte PA (1995) Scientific and ethical aspects of genetic screening of workers: the case of the N-acetyltranferase phenotype. J Clin Epidemiol, **48**: 189-197.

Vineis P, Caporaso N, Tannenbaum SR, Skipper PL, Glogowski J, Bartsch H, Coda M, Talaska G, & Kadlubar F (1990) Acetylation phenotype, carcinogen-hemoglobin adducts, and cigarette smoking. Cancer Res, **50**: 3002-3004.

Waters MD & Nolan C (1995) EC/US workshop report: assessment of genetic risks associated with exposure to ethylene oxide, acrylamide, 1,3-butadiene and cyclophosphamide. Mutat Res, **330**: 1-11.

Waters MD, Stack HF, & Jackson MA (1999) Genetic toxicology data in the evaluation of potential human environmental carcinogens. Mutat Res, **437**: 21-49.

Weinberg (1993) Toward a clearer definition of confounding. Am J Epidemiol, **137**: 1-8.

White E (1997) Effects of biomarker measurement error on epidemiological studies. In: Toniolo P, Boffetta P, Shuler DEG, Rothman N, Hulka B, & Pearce N ed. Application of biomarkers in cancer epidemiology. IARC Scientific Publications No. 142, Lyon, International Agency for Research on Cancer, pp 73-93.

WHO (1975) Early detection of health impairment in occupational exposure to health hazards: report of a WHO study group. WHO Technical Report No. 571. Geneva, World Health Organization.

Willett W (1998) Nutritional Epidemiology (2nd Edition) New York, Oxford University Press, pp 484-496.

BIOMARKERS OF EXPOSURE AND EFFECT FOR CARCINOGENICITY

Kari Hemminki

Department of Biosciences at Novum, Karolinska Institute,
141 57 Huddinge, Sweden

CONTENTS

I.1. INTRODUCTION

Biomarkers are often divided into those measuring exposure, effect and susceptibility (Albertini et al., 1996). The division is somewhat ambiguous but serves in the grouping of biomarkers, as shown in Fig. 5. Biomarkers of exposure and effect for carcinogenicity (genotoxic carcinogens), to be covered in this Appendix, include DNA and proteins adducts, cytogenetic changes and point mutations. As shown in Fig. 5, the chapter covers "Biologically effective dose" and "Early genotoxic effects". The biomarkers were chosen because they are 1) beyond measurement of the chemical or its metabolite (internal dose, biological monitoring), 2) currently applicable to humans, 3) detectable in healthy individuals, unrelated to pre-cancerous conditions (early diagnosis of cancer is outside the scope of this chapter), and 4) mechanistically related to cancer (however for protein adducts the link is indirect). Validity aspects of biomarker applications have been discussed in Appendix 4 and separate publications (IARC, 1997), and this Appendix is mainly limited to technical validity (performance of the assay for human specimens) rather than predictivity for cancer outcome.

Cytogenetic tests involving scoring of microscopic chromosomal aberrations are the oldest of biomarkers used and are still applied in, for example, accidental radiation exposure. Subsequently, sister chromatid changes and micronuclei have been introduced. Recently, *in situ* fluorescence techniques (FISH) have been used in order to score specific chromosomes and chromosomal loci. Applications of DNA and protein adducts and point mutations in human biomonitoring are also recent, most publications having appeared during the past decade.

There have been substantial technical developments in methods of biomonitoring. Here techniques will be discussed only as far as is necessary for the interpretation of the results. For DNA adducts the main techniques, including the ^{32}P-postlabelling technique, will be described. For protein adducts, haemoglobin adduct techniques will be presented. For point mutations, only two of many systems, i.e., those based on the hypoxanthine-guanine phosphoribosyl transferase (HPRT) gene and on the glycophorin A gene, are discussed here. The modern rather than the conventional cytogenetic techniques will be discussed.

Internal dose ⟶ Biologically effective dose ⟶

Biomarkers of exposure

| Carcinogens/mutagens
Metabolites
Blood
Urine
Faeces
Tissues | Protein
adducts

Albumin
Haemoglobin | DNA
adducts

WBC
Urine
Tissue | SCE |

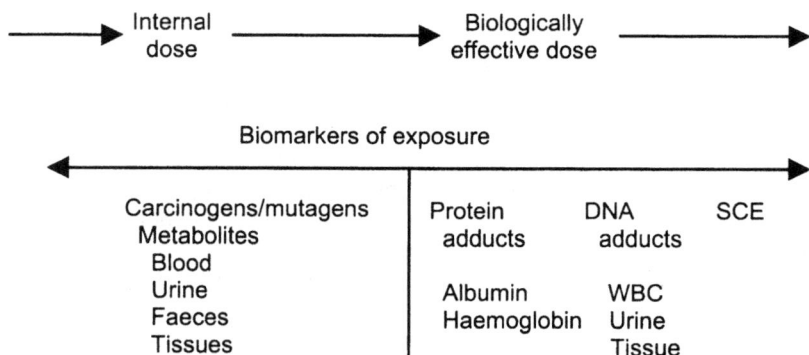

Fig. 5a. Biomarkers of exposure
Abbreviations: SCE = sister chromatid exchanges; WBC = white blood

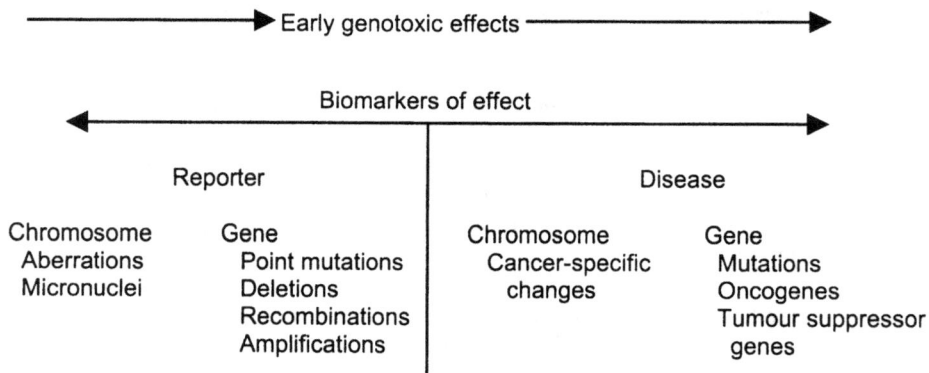

Early genotoxic effects

Biomarkers of effect

Reporter		Disease	
Chromosome Aberrations Micronuclei	Gene Point mutations Deletions Recombinations Amplifications	Chromosome Cancer-specific changes	Gene Mutations Oncogenes Tumour suppressor genes

Fig. 5b. Biomarkers of effect

There is an apparent imbalance in the length of text devoted to the different biomarkers. However, the length is probably roughly proportional to the volume of papers published applying these biomarkers in human studies from occupational and environmental settings. DNA adducts weigh heavily in such comparisons, while the biomonitoring field of point mutations and fluorescence analysis of chromosomal aberrations has been practised by relatively few research groups so far.

I.2. MECHANISMS OF CARCINOGENESIS

Carcinogens are usually mutagens and result in DNA adducts (Harris, 1996; Bartsch, 1996; Loechler, 1996; Nestmann et al., 1996; Ottender & Lutz, 1999). The mechanisms of adduct-induced mutagenesis are illustrated in site-specific mutagenesis studies in which specific adducts are built in a vector and allowed to replicate in a bacterial or eukaryotic host for the scoring of mutations (Loechler, 1996; Dogliotti, 1996; Verghis et al., 1997; Lavrukhin & Lloyd, 1998; Shibutani et al., 1998; Ponten et al., 1999; Tareshima et al., 1999). Studies have also focused on specific adducts, affected nucleotide sites and mutations (Pfeifer & Denissenko, 1998; Ross & Nesnow, 1999). Mutational spectra in tumours have been used as a means of deducing possible causative agents (Harris, 1996; Brauch et al., 1999; Vineis et al., 1999). The conclusions from these studies are that almost all adducts do cause mutations but misreplication is a relatively rare event. Thus the mechanistic chain from adducts to cancer is characterized by many stochastic processes: 1) location of the adduct in the coding DNA, 2) DNA repair, 3) probability of mutation in DNA replication or error-prone repair, 4) mutation in the critical genes, 5) mutation in both alleles of a tumour suppressor gene, 6) accumulation of these mutation in a stem cell and 7) induction of clonal growth (Loeb, 1994). Measurement of a DNA adduct in the bulk DNA is not directly informative of these processes, but an increase in adduct levels in bulk DNA increases the probability of mutations in the critical genes. This is a corollary of the mass law, which does operate in living systems as well.

A sporadic adduct determination is of course rather non-informative if not related to consideration of the duration of exposure. If a particular exposure has lasted for an extended period of time or if it has been excessive, such as anticancer chemotherapy or an accident with radioactive material, some increase in risk may ensue. Although initially emphasized in the "initiation" phase of carcinogenesis, recent evidence on genetic lesions in multiple steps of cancer development suggests that DNA damage has a role in many stages of oncogenesis (Herrero-Jimez et al., 2000).

The role of mutations has not been questioned to the same extent as that of DNA adducts because mutations in growth-controlling genes play an important role in the development of cancer (Harris, 1996; Vineis et al., 1999; Herrero-Jimez et al., 2000). However the appearance of mutations in surrogate tissues of healthy individuals may not be directly informative of the events in target tissues. Additionally, all the mutational systems available for human biomonitoring have their own features and limitations (Cole & Skopek, 1994; Albertini et al., 1996; Tates & Lambert, 1999). Unselected mutations in human target tissues have been measured in mitochondia of which hundreds of copies are present in each cell (Coller et al., 1998).

Many types of chromosomal changes are related to cancer, including for example the specific translocation, Philadelphia chromosome, detected in most cases of chronic myeloid leukaemia. However, the role of non-specific chromosomal aberrations has been less clear until recently. In two follow-up studies, the relationship to cancer has been directly strengthened (Hagmar et al., 1994; Bonassi et al., 1995). The subjects who historically had an increased level of chromosomal aberrations were found to be likely to develop cancer more often than those registered with low levels of chromosomal aberrations. Sister chromatid exchange rates were not related to the risk of cancer, while the data for micronuclei were too sparse to allow conclusions. These results have been expanded, and it has been shown that the increase in chromosomal aberration frequence is unrelated to exposure (Bonassi et al., 2000).

Relevant to DNA adducts is that many DNA repair deficiency syndromes involve vast increases in risk of cancer. Xeroderma pigmentosum patients experience some 1000-fold increase in the incidence of non-melanoma skin cancer and melanoma (MacKie,

1996). Hereditary non-polyposis colorectal cancer (HNPCC), due to germline DNA mismatch repair gene mutations, predispose affected individuals to a 70-fold risk of colorectal and endometrial cancer (Aarnio et al., 1999). This syndrome is characterized by genetic instability at simple DNA repeat sequences because of DNA mismatch repair deficiency (Dunlop et al., 1997). *P53* has a role in arresting the cell cycle after DNA damage and thus allowing time for DNA repair; a germline mutation in the Li-Fraumeni families is consonant with increased risks of many cancers (Malkin, 1998). Bloom syndrome patients have a deficient DNA helicase (German & Ellis, 1998), and even the breast cancer genes, *BRCA 1* and *BRCA2* are suspected of being involved in DNA repair or maintenance of genome integrity.

Mechanistically based risk assessment has been advocated recently, as highlighted in preambles to IARC monographs (IARC, 1996). The biomarkers discussed here provide a basis for such a risk estimation and the availability of human data makes them very relevant (Törnqvist & Ehrenberg, 1994; Granath et al., 1999; van Sittert et al., 2000).

I.3. DNA ADDUCTS: METHODS AND LIMITATIONS

The methods used in the determination of DNA adducts in humans are listed in Table 7. The usefulness of a method in human biomonitoring requires high sensitivity because the levels of adducts are low. The methods used are divided in Table 7 in those generally applicable to most types of DNA adducts and those applicable to certain classes only. The methodology for determination of DNA adducts is under constant development and diversification. Only the most commonly used methods are discussed in some detail below. More extensive reviews can be found elsewhere (IARC, 1993, 1994, 1997). Some comments are made about the sensitivity of the methods. This tends to depend much on the compound or chemical class studied. Additionally, any prepurification of the adduct helps to boost sensitivity but the amount of DNA may become limiting. The DNA in one diploid human cell contains 6×10^9 nucleotides, i.e.,

Table 7. Methods for the detection of DNA adducts in humans

Method	Advantages	Disadvantages	Technical validity
For most adducts			
^{32}P-postlabelling	very sensitive, small amounts of DNA	laborious, radiation	valid when standards used
Immunoassays	sensitive, easy, preparative columns	raising antibodies, specificity	questionable as primary assay
GC-MS	specific, quantitative	cost, volatilization	valid when standards used
For certain adducts			
Electrochemical detection	easy, sensitive, cheap	specificity, contaminants	valid, risk for contamination
Fluorescence	easy, specific	large amounts of DNA, contaminants	valid, risk for contamination
Alkyltransferase	specific class of adducts, cheap, easy	specific O^6-alkyl-guanines ?	valid
Atomic Absorption	specific, sensitive	specific metals, Pt	valid

6 pg of DNA. As human blood contains some 7×10^6 white blood cell/ml, 10 ml of blood can yield up to 400 µg of DNA.

The ^{32}P-postlabelling assay, described in the early 1980s (IARC, 1993), is a sensitive technique that is now extensively used in human biomonitoring. The sensitivity of the technique depends on the use of high specific activity ^{32}P-ATP in the kinase reaction, where a radioactive phosphate group is transferred to 3'-nucleotides. The method improved in specificity and cost when nuclease P1, butanol, high-performance liquid chromatography (HPLC) and immunoaffinity chromatography techniques were used to prepurify adducts before or after the kinase reaction. The advantages of this method are sensitivity, small samples size and reasonable start-up costs. The detection limits are in the order of 1 adduct in 10^9 normal nucleotides with 5 μg of DNA. The method involves many steps, which may reduce its reproducibility. Typically, the reported levels of adducts vary between laboratories by at least one order of magnitude. An international interlaboratory study revealed, among other things, a need for harmonization of the assay protocols (IARC, 1994). However, as some five different enzymes, ATP and thin-layer plates, many of which vary by batch, are used in postlabelling, no cook-book standard protocol can be laid out. Unfortunately, commercial products too often fail to meet the manufacturer's specifications.

In order to provide a quantitative assay, standard compounds have been used in the identification and quantification of adducts. Although this is widely accepted now, it was long thought by a large section of the postlabelling community that labelling of all adducts was complete. It has now been demonstrated with tens of different synthetic postlabelling standards that, depending on the adducts and conditions of labelling, the recoveries vary between 0 and 100%. Even diastereomers can label differently. In an illustrative experiment, DNA adducts of a number of ^{3}H-labelled polycyclic aromatic hydrocarbons (PAHs) were prepared in a microsomal system and used for optimization and measurements of recoveries in the postlabelling assay (Segerbäck & Vodicka, 1993). The optimal labelling conditions for all tested compounds were very similar. The recoveries varied from 3 to 60% among different PAHs, indicating that the levels of these adducts could be considerably underestimated when analysing human samples from PAH-exposed populations. As similar results have been reported with entirely different groups of compounds, it can be generally concluded that optimal labelling conditions are adduct-specific (Hemminki et al., 1991a,b). Thus the absence of proper standards, or analysis of unknown adducts, impedes quantitative interpretation of postlabelling results. Extensive surveys of the literature are available (IARC, 1993, 1994).

High-performance liquid chromatography (HPLC) has become increasingly useful in the analysis of radioactive products from postlabelling reactions. The yields are high and the level of separation and reproducibility are much improved over those of thin-layer chromatography (TLC) (Hemminki et al., 1996, 1997a,b; Zhao et al., 1999).

The results of an interlaboratory trial between many laboratories familiar with the postlabelling technique showed coefficients of variation (SD/mean) ranging between 35 and 75% between the different laboratories depending on the type of adduct assayed (Castegnaro & Phillips, 1997). Standard compounds were available in the analysis which allowed for normalization of results. The method is so complex that the use of some standards as day-by-day indicators of the performance of the assay is mandatory for quantitative analysis. In another interlaboratory comparison samples from ethene-exposed rats were analysed by the postlabelling assay in one laboratory and MS in another, with excellent consistency (Eide et al., 1999).

The continued application of poorly standardizable TLC methods for analysis of unknown adducts has given a notorious reputation to the postlabelling technique. However, this is entirely unjustified because the method can be used as a specific and reproducable analytical technique when synthetic standard compounds and HPLC analysis are being applied, as discussed above (Eide et al., 1999). This technique has been used to quantify a number of specific human DNA adducts (Zhao et al., 1999, 2000; Plna et al., 2000).

During the most recent decade, highly specific polyclonal and monoclonal antibodies have been developed for detection of DNA adducts formed by carcinogens such as by PAHs, aromatic amines, mycotoxins, aldehydes, alkylating agents, chemotherapeutic agents and ultraviolet light (Strickland et al., 1993; Hemminki et al., 1995; Clingen et al., 1995). The antigens were DNA, nucleotides or nucleosides reacted with the carcinogen. Both polyclonal and monoclonal antibodies were produced. Immunoassays have been carried out as radioimmunoassay (RIA), enzyme-linked immuno-sorbent assay (ELISA) and ultrasensitive enzyme radioimmunoassay (USERIA). Sensitivity depends both on the antibody used (i.e. affinity constant) and the type of immunoassay applied. The secondary antibody is conjugated to an enzyme that catalyses a reaction forming a radioactive (USERIA), coloured or fluorescent (ELISA) product. The assays are generally sensitive enough to detect one adduct in 10^8 nucleotides in 50 μg DNA (Santella et al., 1993). Assay specificity depends upon the cross-reactivity of the antibody with DNA or other DNA adducts. For instance, antibodies made against benzo(a)pyrene (BP)-DNA adducts cross-react with related adducts of chrysene and benz(a)anthracene. Such a cross-reactivity towards closely-related adducts can be considered an advantage for class-specific detection but a disadvantage for compound-specific detection (Strickland et al., 1993). Raising of specific antibodies is time-consuming but once developed the assay is easy and inexpensive. Quantification is done by comparing to *in vitro* modified standards. However, these display different affinities depending on modification levels, which creates uncertainties about the absolute amounts of adducts present, particularly at the low levels detectable in humans.

Many antibodies have been described in the literature towards many different types of adducts with alleged specificity and sensitivity (Strickland et al., 1993; Poirier, 1997). However, applications for human samples have been limited. Thus it may not be advisable to use DNA adduct antibodies as a quantitative assay but rather as a purification step before the quantitative assay, such as postlabelling (Bartsch, 1996).

Mass spectrometry (MS) coupled to gas chromatography (GC) or other interphases is gaining increasing ground in the analysis of DNA adducts (Allam et al., 1990; Bakthavachalam et al., 1990; Chaudhary et al., 1994, 1995). The technique is powerful but the instrumentation is expensive. Volatilization of the samples has been one of the main obstacles so far. In its present form the technique appears most suitable to relatively small and abundant adducts that are analysed as base derivatives, an area gaining increasing application (Zhao et al., 1999).

The remaining techniques listed in Table 7, including fluorescence spectroscopy, electrochemical detection, O^6-alkyl-transferase assay and atomic absorption spectroscopy, are applicable to a limited number of adducts but may be valuable for those (Hemminki, 1995). For example, synchronous fluorescence spectroscopy can be applied in the identification of individual fluorescent species from a mixture of similar compounds. Fluorescence detection systems are quite sensitive and they have been successfully applied to the analysis of BP tetrols, released hydrolytically from DNA (Alexandrov et al., 1992).

I.3.1 DNA adduct in humans

I.3.1.1 Assay of complex polyaromatic adducts

Human exposure to PAHs involves numerous different congeners and related compound such as nitro-PAH and heterocyclic compounds. In spite of the problems in the interpretation of postlabelling results of complex mixtures, most published articles concern exposures where PAHs are of primary concern. Many of the groups studied have been at a risk of cancer according to epidemiological studies that reflect exposures a few decades ago. The main questions posed have been: 1) do the exposed groups show higher, exposure-related adduct levels than the controls; 2) is there a correlation between exposure measures (air concentration or urinary 1-hydroxypyrene) and adducts; 3) how large are the interindividual variations and 4) what are the effects of metabolic genotypes on the level of adducts.

There has been uncertainty about the levels of PAH-like adducts in human specimens relating to the techniques used. DNA adducts of PAHs appear to be present in most human tissues, as shown by four independent techniques available for PAH-DNA adduct analysis: ^{32}P-postlabelling, immunoassay, HPLC-fluorescence detection and synchronous fluorescence spectroscopy (Rojas et al., 1994, 1995). The techniques have different specificities to detect individual species of PAHs, such as BP, usually constituting 1–20% of the particulate PAH. The apparent levels of PAH-DNA adducts in human tissues differ vastly depending on the method used. For instance, in white blood cells the postlabelling assay applying thin-layer chromatography (TLC) usually yields total PAH adduct levels lower than BP-DNA adduct levels alone by the other techniques (Hemminki et al., 1995).

In the previous postlabelling analyses total PAH adducts in smoker's lungs have been noted at 1–30 adducts/10^8 nucleotides (Beach & Gupta, 1992). Immunoassays using BP-DNA antibodies have recorded levels ranging between 1 and 100 adducts/10^8

nucleotides (Hemminki et al., 1995). It remains unclear what proportion of this would be BP because the antibodies cross-react with adducts of several PAHs. The HPLC-fluorescence detection data are still limited, the range being 1–10 BP diol epoxide adducts/10^8 nucleotides, depending on the metabolic genotype (Alexandrov et al., 1992; Bartsch, 1996). Incidentally, these HPLC-fluorescence detection data correlated with the postlabelling data when the TLC area of BP adduct was analysed. The postlabelling-HPLC results on total PAH adducts range from 17 to 230 adducts/10^8 nucleotides (Hemminki et al., 1997b). Considering that BP adducts are a minor proportion of all lung PAH-like adducts, these postlabelling data are quite compatible with those obtained by HPLC-fluorescence detection, and are probably reasonable estimates of the total PAH-like adduct levels in human lung DNA.

I.3.1.2   *Aromatic DNA adducts in occupational and environmental exposure*

Study populations have included foundry and coke workers, aluminium and electrode workers and chimney sweeps. Several studied on foundry workers have shown elevated total white blood cell DNA adduct levels measured by immunoassay (Perera et al., 1992; Santella et al., 1993) and postlabelling (Perera et al., 1993, 1994), relating to exposure. Among the other occupational groups, coke workers had higher levels of aromatic adducts than the local controls (Grzybowska et al., 1993; Widlak et al., 1996). Somewhat elevated but not statistically significant differences were seen in electrode and aluminium workers, even though air concentrations of PAHs and urinary 1-hydroxypyrene levels indicated excessive exposure (Ovrebo et al., 1994, 1995). Adduct levels were also slightly increased in total white blood cell DNA of chimney sweeps but the difference to a control group became significant only after adjustment for the CYP1A1 and glutathione transferase (GSTM1) genotype (Ichiba et al., 1994). In all the studies cited the inter-individual variation in the levels of adducts has been large, over 10-fold. The variation is usually larger in the exposed as compared to the control populations, suggesting that exposures as well as constitutional factors contribute to such a variation.

Some occupational groups exposed to car and diesel exhaust have been studied. Diesel exhaust contains PAHs and nitro-PAHs, the latter of which may be metabolized to aromatic amines and further acetylated (Hou et al., 1995). The groups included garage workers, overhauling diesel buses and inhaling diesel exhaust gases, car mechanics exposed to spilled engine oils but not to exhausts, and truck terminal workers, unloading and reloading diesel trucks (Hemminki et al., 1994a). All these groups had increased levels of lymphocyte DNA adducts, the highest levels being correlated with air concentrations of diesel exhausts. The estimated air BP levels were below 10 ng/m^3. GSTM1 and *N*-acetyl transferase (NAT2) genotypes were determined in the study subjects. In the individuals with a combined genotype of slow acetylation, lacking the GSTM1 gene, the adduct levels were significantly increased (Hou et al., 1995). Neither genotype alone had an effect on the level of adducts.

Some DNA adduct studies have examined the problem of urban air pollution in focusing on populations that stay most of their time at street level in urban centers, such as bus drivers and newspaper vendors. In one study bus drivers from central Stockholm and from the outskirts were compared to a non-occupational control group, fine mechanics. All the participants were non-smokers. PAH-type aromatic DNA adducts in lymphocytes, PAH adducts in albumin and ethene and propene adducts in haemoglobin were not elevated in the urban bus drivers (Hemminki et al., 1994b). A similar type of study was carried out in Copenhagen, where bus drivers had clearly increased aromatic DNA adduct levels, analysed by the TLC method (Nielsen et al., 1996; Knudsen et al., 1999). The effects of metabolic genotypes on the level of adducts was also studied but showed no significant effect of GSTM1 or NAT2. In Milan the study subjects were newspaper vendors from busy streets and the outskirts of Milan. The levels of aromatic DNA adducts, assayed for by the TLC method, did not differ in these populations.

A series of environmental studies have been carried in Silesia, a heavily industrialized area of Poland, initiated several years ago in response to alarming reports of environmental pollution. The first study on the Silesian population showed an elevated level of adducts, by postlabelling and immunoassay, in the total white blood cells of the residents (Hemminki et al., 1990a). This was followed by reports on seasonal differences in adduct levels, which matched the air concentrations of PAHs. The effects were mainly seen in the DNA of the long-lived lymphocytes, while granulocytes showed no clear effect. Sampling in the summer and winter allowed a rough estimation of the half-lives of aromatic adducts in lymphocytes to be 1–2 months (Grzybowska et al., 1993). In addition, cytogenetic damage was seen in the Silesian population (Perera et al., 1992).

The nature of the adducts detected by postlabelling has been studied in more detail by comparing nuclease P1, butanol extraction and immunoaffinity purification of the adducts. The adduct recovery was approximately equal by the P1 and butanol techniques, suggesting that the adducts are of the PAH type. For immunoaffinity chromatography, an antibody raised against BP diol epoxide-DNA was used. Only about 25% of the adducts were bound by the antibody, indicating that most of the adducts in DNA are not closely related to BP. However, in winter time of high air pollution, the relative binding by the immunoaffinity column was higher than in the summer time (Widlak et al., 1996). Using postlabelling-HPLC analysis with an on-line flow-through radioactivity detector, typical seasonal adduct peaks were noted and they were particularly prominent in lymphocyte DNA collected in the winter time. They eluted in the area of PAH-DNA adducts, giving additional support that the adducts are PAH-like (Möller et al., 1996).

One of the enigmas of DNA adduct studies concerning complex PAHs has been the difficulty of observing dose-response relationships (Hemminki et al., 1997c). While dose-response relationships have been observed in leucocyte adducts of foundry workers, exposed up to $50 \, ng/m^3$ BP, the level found in environmental air samples from polluted regions of Poland as discussed above, involving massive PAH exposures ($> 1 \, \mu g/m^3$) in aluminium, coke and electrode industries, have barely caused increases in adduct levels (Hemminki et al., 1997c). A recent study on dose-response relationships in smokers' lymphocytes appears to offer an explanation to this enigma (Dallinga et al., 1998). The authors observed that aromatic adducts saturated at a level of 10–15 cigarettes/day. The authors also measured 4-aminobiphenyl-haemoglobin adducts and noted saturation at the same level of smoking, making the results very persuasive. It appears that the suggested saturation levels are not very far from those that can apparently be reached by a rich dietary intake of PAHs (van Maanen et al., 1994; Hemminki et al., 1997c; Poirier, 1997).

61

I.3.1.3   Alkenes

The main technical consideration for small adducts, such as those caused by alkenes, is that purification from unmodified nucleotides has to be carried out prior to labelling to guarantee good labelling efficiency. Different chromatographic techniques can be used because nuclease P1 and butanol extraction, applicable for PAH, may cause loss of the adducts. Among alkenes, DNA adducts of ethene have been measured in humans by the postlabelling assay and correlated to smoking (Kumar & Hemminki 1996; Zhao et al., 1999). This assay would equally detect 7-alkylguanine derivatives of propene, butadiene (Kumar et al., 1995, 1996) and styrene (Hemminki & Vodicka, 1995). Human DNA adducts of butadiene and epichorohydrin, both important industrial chemicals, have been identified in production workers (Zhao et al., 2000; Plna et al., 2000). The postlabelling assay for styrene, based on quantification of O^6-guanine adducts, has been applied to lamination workers in a series of studies (Vodicka et al., 1993, 1994, 1995). The same lamination workers were sampled periodically in order to measure lymphocyte and granulocyte DNA. Again it was shown that high DNA adduct levels were found in lymphocytes. The repair of O^6-adducts appeared to be slow as no essential decrease in adduct levels was noted after a two-week vacation. Styrene-haemoglobin adducts and DNA strand-breaks, measured by the comet assay, were also increased in the lamination workers. This was the first time a correlation was observed in DNA and haemoglobin adducts induced by a specific agent in humans.

The genotoxicity of styrene has been of interest world-wide because it is one of the few suspected mutagenic compounds that may cause daily exposures in gram quantities (Hemminki & Vodicka, 1995). Styrene is an example of how the chemical characterization leads to production of standard compounds for postlabelling and to a quantification of O^6-guanine adducts in white blood cell DNA of lamination workers. Further studies have been carried out to sample the same lamination workers periodically in order to measure lymphocyte and granulocyte DNA. It was shown that adducts are essentially only found in lymphocytes. The repair of O^6-adducts appeared to be slow as no essential decrease in adduct levels was noted after a two-week vacation (Vodicka et al., 1994).

The adduct studies suggested that the factory controls in fact are not completely unexposed. Strand-breaks, measured by the comet assay, were also increased in the lamination workers. There was a correlation between strand breaks and O^6-guanine adducts, but neither correlated with HPRT mutant frequency (Vodicka et al., 1995; Somorovska et al., 1999). *In vitro* data on the effect of styrene oxide, the main metabolite of styrene, on cultured human lymphocytes confirmed the relatively long half-lives of O^6-guanine DNA adducts and the induction of strand breaks (Bastlova et al., 1995).

I.3.1.4   UV-induced damage

UV-induced DNA damage is important when validity aspects of biomarkers are considered: 1) the doses can be well controlled and it is ethically acceptable to expose normal humans to doses that inflict a minimal erythemal response (MED) of about 200 J/m^2 in a fair-skinned population; 2) UV irradiation induced specific types of DNA damage, cross-linking two adjacent pyrimidines to a cyclobutane dimer or to a 6–4 photoproduct (Bykov et al., 1998a,b); 3) these adducts induce specific tandem CC to TT mutations, also in defined genes from skin tumours (Harris, 1996); 4) UV irradiation is a well-known cause of human cancer; 5) target tissue is available for experimental study.

For DNA adduct analysis, a modification of the postlabelling technique was necessary because it was found that cross-linked dinucleotides label very poorly (Bykov et al., 1995). In the modification, a normal nucleotide was left on the 5'-side of the cross-linked dinucleotide, resulting in a number of labelled trinucleotides. This kind of modification was so far the only way to label cross-linked products. Radioactivity was analysed by HPLC radioactivity detection and the products were assigned based on the standards used (Bykov & Hemminki, 1996). The assay has been used to study dose-response relationships in humans. Skin biopsies from UV-irradiated skin demonstrated a linear dose-response between dose from 50 to 400 J/m^2 and adducts (Fig. 6, Bykov et al., 1998a). The dose-relationship was also linear between 150 and 2000 J/m^2 when a sunscreen was used to protect skin, indeed reducing the level of adducts (Bykov et al., 1998b). An unexpected finding in these studies was a large, 30-fold difference between individuals of similar skin type in the immediate level of adducts.

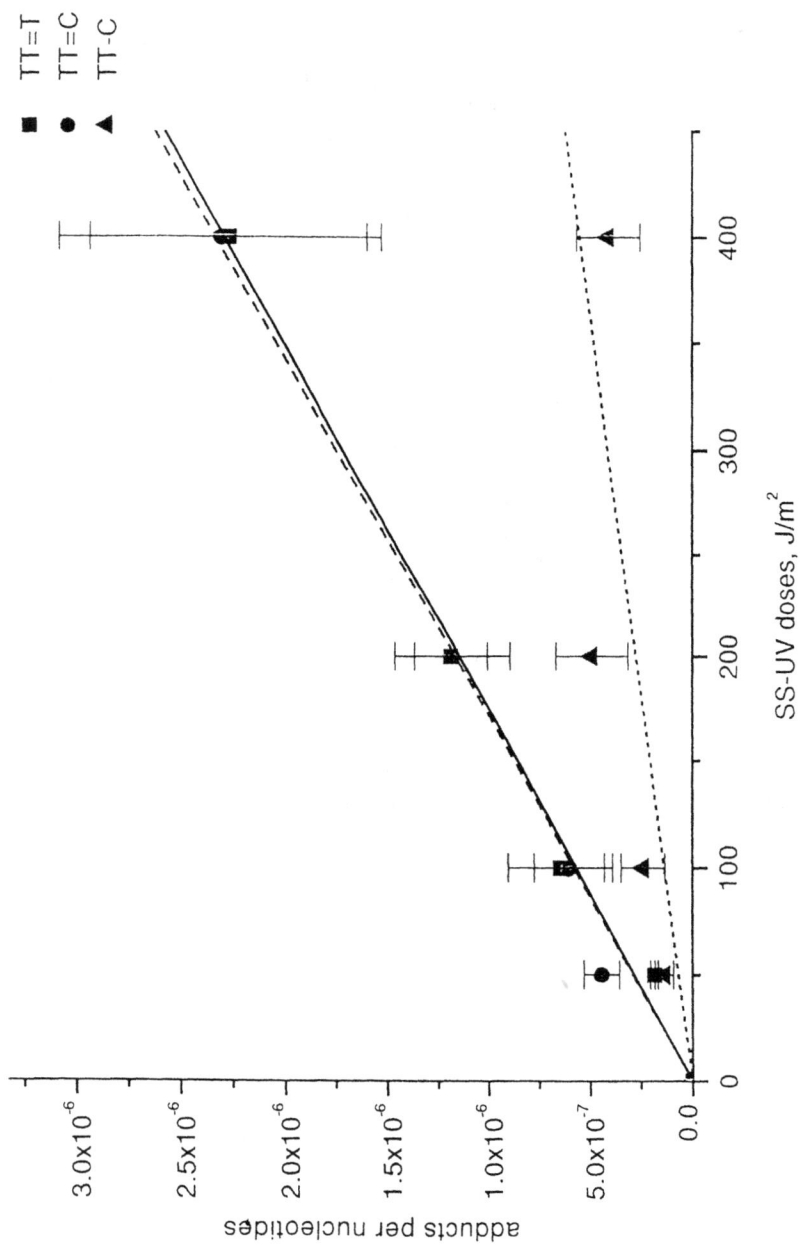

Fig. 6. Photoproducts induction, average values

Solar UV-irradiation is thought to be an important cause of non-melanomatous skin cancer but it may also contribute to melanoma. Mammalian cells can repair the adducts at various rates (Clingen et al., 1995; Young et al., 1996). The relationship between DNA repair and cancer is illustrated by several skin diseases, such as xeroderma pigmentosum, where a repair defect predisposes to skin cancer. Decreased repair of UV damage may also contribute to common skin tumours such as basal cell carcinoma. Specific UV-induced photoproducts may be measured by the ^{32}P-postlabelling technique in humans exposed experimentally and through sun tanning; DNA repair efficiency can be determined and related to skin diseases (Hemminki et al., 1999; Bykov et al., 1999; Xu et al., 2000a,b).

I.3.1.5   *Adducts excreted in urine*

The application of urine as a source of adducts for biomonitoring has been reviewed (Shuker & Farmer, 1992). The advantages of using urinary adducts as the measure of exposure to genotoxic compounds include non-invasiveness, large amounts of adducts available and at least in some cases relatively simple purification of adducts. The main disadvantage is that the molecular origin of the adducts is unknown, whether it is DNA, RNA or free nucleosides. Owing to abundance, cytoplasmic location and fast turn-over, RNA adducts are likely to overwhelm DNA adducts unless thymine adducts are investigated. The tissue source is also unknown and dietary sources may confound the results. If dietary sources can be excluded, urinary adducts represent an integrated measure of whole body dose for carcinogens. However, kinetic parameters of DNA adduct formation and repair in human tissues *in vivo* are almost completely unknown and any quantitative assessment of data will be arbitrary.

The methods are similar to those used for DNA adducts, except that ^{32}P-postlabelling cannot be used. Immunoassay, GC-MS and HPLC-electrochemical detection have been the main techniques (Shuker & Farmer, 1992). A number of adducts have been studied in humans, including alkylation products of many bases (Prevost et al., 1996), oxidation products, aflatoxin and BP adducts. The assumed measure of oxidative damage, 8-hydroxydeoxyguanosine, has been correlated to many chemical and physical exposures, and physiological and pathological processes (Shigenaga et al., 1994; Tagesson et al., 1995).

The major DNA adduct formed by the carcinogenic mycotoxin aflatoxin B1 (AFB) has been identified as aflatoxin-N7-guanine. A positive correlation between the dietary intake of AFB and the level of urinary AFB-guanine has been established in both a West African and a Chinese population using immunoaffinity chromatography in combination with HPLC to isolate and quantify AFB-guanine (Groopman et al., 1992a,b, 1993).

A follow-up study has been carried out in the Shanghai region, an area with a high incidence of liver cancer (Qian et al., 1994). A cohort of 18 000 men was recruited and interviewed in 1989. A urine sample was collected and stored for a nested case-control analysis. After 58 000 person years, 50 liver cancers have been identified and compared to 267 matched controls from the cohort. About half of the individuals had detectable levels of AFB metabolites or 7-AFB-guanines in urine. The presence of AFB derivatives in urine confirmed a liver cancer risk of 3.5, of hepatitis B virus surface antigens a risk of 7.9, and of both markers a risk of 57. A weakness is that only one urine sample was collected, giving a cross-sectional, short-term exposure assessment.

I.3.1.6   *Adducts and metabolic genotypes*

In addition to being indicators for exposure, DNA adducts would appear excellent tools in carcinogen metabolism and DNA repair work. However, this area has been underdeveloped because the methods for adduct determination have not been available for specific adducts in humans. Large amounts of work have been devoted to study relationships between metabolic phenotypes/ genotypes and cancer, with an implied assumption that DNA-damaging metabolites are involved. The results have been a bewildering collection of contradictory data. However, recently, when specific DNA methods have been applied, coherent results on the effects of certain metabolic genotypes have begun to emerge (Bartsch et al., 1998; Hou et al., 1999; Thompson et al., 1999; Lunn et al., 1999). More details are presented in Appendix 3.

I.4.   HAEMOGLOBIN ADDUCTS

Haemoglobin adducts are excellent monitors of the absorbed and metabolically activated dose of the chemical in the body.

Relatively large amounts of haemoglobin are available in human blood (140 mg/ml). Because the life-time of haemoglobin is 4 months, it measures the exposure over a relatively long time period (Neumann, 1984; Törnqvist & Landin, 1995). Protein adducts are also measured in albumin but the techniques are essentially similar to those used for haemoglobin (Ferreira et al., 1994; Tas et al., 1994).

Among the amino acids targets in haemoglobin, cysteine, histidine, methionine, glutamic acid, aspartic acid and *N*-terminal valine form adducts. The methods applied either release the adducted amino acid or cause hydrolysis of the ligand to a free form. The products are then analysed directly or after derivatization. Bulky adducts can be readily analysed as hydrolysis products. For alkylation products a specific technique has successfully been used. In this the Edman degradation technique, developed for protein sequencing, was modified for the determination of *N*-terminal valine adducts. The products were analysed by mass spectrometry. This technique is now used by a number of groups in the world and has enabled wide application in human biomonitoring (Törnqvist & Landin, 1995).

The kinetics of haemoglobin adducts are well established and are much easier than those of DNA. Unlike the repair systems for DNA adducts, no repair enzymes are known for carcinogen-modified proteins. Thus, haemoglobin adducts turn over with the protein. Most known haemoglobin adducts arise through the same metabolic pathway that leads to DNA adducts. The levels of adducts in haemoglobin and DNA are thus likely to correlate, although the evidence on this in humans is meagre (Törnqvist & Landin, 1995; Granath et al., 1999; van Sittert et al., 2000).

Because of the GC-MS techniques applied, specific adducts can be determined and using such methods it was established that there are adducts in humans and experimental animals originating from endogenous sources. Quantitative data are available on the level of alkylation and aldehyde products in human DNA. For example, adduct formation by ethene/ethylene oxide and other alkenes has been extensively studied in humans using Edman degradation with

GC-MS. In non-smokers background levels of *N*-terminal valine adducts of ethene, propene, butadiene and styrene (or the corresponding epoxides) are 20, 2, < 0.1 and < 0.1 pmol/g haemoglobin, respectively (Table 8, Osterman-Golkar & Bond, 1996; Osterman-Golkar et al., 1996). In addition to endogenous production, passive smoking and urban air may be sources of the background alkylation. Dose-response relationships have been investigated in smokers and workers with occupational exposures. Smoking increases ethene- and propene-related adducts by 9 and 0.2 pmol/g haemoglobin/cigarette per day, respectively. Occupational exposure at 1 mg/m^3 for 40 h/week increases the adducts for ethene and propene approximately by 120 and 10 pmol/g haemoglobin, respectively (Osterman-Golkar & Bond, 1996). The detection limit is approximately 0.05 pmol/g haemoglobin, which would, in principle, allow detection of ethene and propene adducts at environmental exposures. Haemoglobin adducts have been measured among many other end-points in workers exposed to styrene and butadiene (Somorovska et al., 1999; Hayes et al., 2000). Inter-laboratory comparisons have been carried using human samples, resulting in some 10-fold differences in results between the laboratories (Törnqvist et al., 1992). However, these could be mainly accounted for by the different adduct standards used, and the ranking of results was consistent within laboratories.

I.5. POINT MUTATIONS

The point mutation systems available for human biomonitoring measure mutant frequency in circulating blood cells as an attempt to estimate mutation frequency (Cole & Skopek, 1994), the difference between these frequencies being the same as with prevalence and incidence rates, respectively, in epidemiology. There are about five systems with some human data, but only the HPRT and glycophorin A assays have been used for biomonitoring purposes. The mutant frequency is measured either by a metabolic selection system (HPRT) or by quantifying the allelic forms of proteins by advanced optical reading systems (glycophorin A) (Albertini et al., 1996). In the T-cell clonal assay for HPRT the selection takes place by culture

Table 8. Haemoglobin (*N*-terminal valine) adduct levels (pmol/g globin) resulting from exposure to butadiene and some other low molecular weight compounds in referents (non-smokers and smokers) and subjects with occupational exposure (Osterman-Golkar & Bond, 1996)

Compound (reactive intermediate)	Background adduct level (average, range or mean ± SD)	Type of exposure	Exposure concentration	Adduct level (average, range)	Binding index (pmol/g globin per ppm × hour unless otherwise stated)[a]
Butadiene (epoxybutene)	≤ 0.13[b]	occupational lab and maintenance work	0.1 (median value)	< 0.1[b]	
		process work	1 ppm (median value)	0.16, < 0.1–0.32[b]	$\sim0.5 \times 10^{-3\ b,c}$
Styrene (styrene 7,8-oxide)	< 10[d]	occupational	75 ppm (average)	28, 15–52	$\sim1 \times 10^{-3\ c}$
Ethylene (ethylene oxide)	20, 12–27 (non-smokers)	occupational	0.3 (0.1–1 ppm, 40 hour/week)	43, 22–65 (non-smokers)	0.2 (0.6–0.06)
	16.1 ± 2.1 (non-smokers)	tobacco smoking	1–25 cigarettes/day	146, 50–355	9 pmol/g globin per cigarette per day

Table 8 (contd.)

Compound (reactive intermediate)	Background adduct level (average, range or mean ± SD)	Type of exposure	Exposure concentration	Adduct level (average, range)	Binding index (pmol/g globin per ppm × hour unless otherwise stated)[a]
Ethylene (contd.)	63 ± 20	tobacco smoking	>15 cigarettes/day	361 ± 107 (smoking women; maternal blood)	
	42 ± 18 (newborn babies' blood)			147 ± 68 (newborn babies' blood)	
Ethylene oxide	14–26 (see also data for ethylene)	occupational	low to 28 ppm/week	84–2070[e]	9
Acrylonitrile	< 2	tobacco smoking	5–20 cigarettes/day	2.2–178	9 pmol/g globin per cigarette per day

[a] Calculated assuming that the steady-state adduct level corresponds to exposure at work (8 hour/day, 5 days/week) during 9 weeks.
[b] Only one of the two regioisomers measured.
[c] Order of magnitude.
[d] Current unpublished studies indicate that the background is below 0.1 pmol/g globin.

70

of lymphocytes in the selection medium containing 6-thioguanine, which is incorporated in DNA of wild type cells, causing cell death. The mutants are scored by comparing to the number of cells cultured without the selection system (Hou et al., 1995; Bastlova et al., 1995). The target size of the HPRT gene is 45 000 nucleotides. The mutation frequency is approximately 5×10^{-6} in healthy adults (Cole & Skopek, 1994). The mutations are point mutations, deletions and insertions. In the glycophorin A system the mutant frequency is over 1×10^{-5} and it scores, additionally, somatic recombination or other events leading to loss of heterozygosity (Cole & Skopek, 1994). Thus among healthy heterozygous individuals (MN), point mutations, deletions and insertions (scored as hemizygosity MO or NO) account for roughly equally high mutant frequency as loss of heterozygosity (MM or NN). A comprehensive special journal issue has recently covered the status of HPRT mutational studies (Tates & Lambert, 1999).

Mutational spectra can also be studied but this requires clonal cell populations. This can be done by cloning cultured cells or by analysing DNA from tumours that normally are clonal. Data bases exist for mutational spectra in the HPRT gene using the clonal assay and in the p53 gene from human tumours (Cariello, 1994; Greenblatt et al., 1994). In the HPRT database 580 *in vivo* mutations were included in 1995, dominated by point mutations (45%) and deletions (40%).

Interpretation of results for HPRT mutation frequency in human biomonitoring studies is not simple because a few physiological background variables are known in this system. These uncertainties include origin of the mutations in stem cells versus circulating lymphocytes and half-lives of the mutant cells. These factors can cause changes in apparent mutation frequency both within and between individuals. Mutant clones arise and need to be controlled for. In comparative studies between two laboratories, inter-individual and intra-individual differences in mutant frequencies were large (Cole & Skopek, 1994). Additionally, change in the mutant frequency is not informative of its cause (which agent) and hence a careful study design is critical for valid conclusions. For example,

71

tobacco smoking and age affect the mutant frequency and need to be considered (Albertini et al., 1996). Because the mutations are determined in a sample of peripheral lymphocytes, in which the mutations are determined *in vitro*, the results are given as "mutant frequency" rather than "mutation frequency" to emphasize the cross-sectional nature of the results.

Mutations in the HPRT gene in human lymphocytes have been studied extensively but few occupational and, as far as is known, no environmental studies in relation to chemical exposures have been carried out (Cole & Skopek, 1994; Curry et al., 1999; Tates & Lambert, 1999). The non-occupational studies include those on smoking-, radiation-, chemotherapy- and disease-induced mutation rates. The limited number of occupational studies partially reflect logistic problems, because cell separation from blood has to be carried out within hours of blood collection, and living cells have to be delivered to the laboratory of analysis. With such constraints, sampling of a reasonable number of individuals in the field is a major undertaking. Moreover, large inter-individual differences and dependence of the mutation rates on age and smoking may discourage attempts to distinguish small differences between the exposed and control populations.

Among the occupational groups studied, workers producing the anticancer agent cyclophosphamide have elevated levels of lymphocyte HPRT mutations (Huttner et al., 1990). Exposures to ethylene oxide (Tates et al., 1991) and styrene/dichloromethane have also caused increases in mutation frequency (Tates et al., 1994). In the studies on lamination workers, discussed above, the HPRT mutation frequency was elevated in workers exposed to styrene, but the increase reached statistical significance only when compared to an external rather than factory control group (Vodicka et al., 1994; Somorovska et al., 1999). Induction of HPRT mutations in cultured human lymphocytes exposed to styrene oxide was considered weak (Bastlova et al., 1995).

Mutant frequencies have also been determined in occupational populations exposed to PAHs. In a study on foundry workers, HPRT correlated with exposure and adduct levels, while glycophorin A NO mutations showed a moderate, but statistically insignificant, trend with exposure (Perera et al., 1993, 1994). The HPRT mutant frequency was not increased in garage workers but at an individual level there was a highly significant correlation between adducts and mutant frequency (r = approximately 0.35). GSTM1 and *N*-acetyl transferase (NAT2) genotypes were determined in the study subjects, but the genotypes, alone or combined, had no effect on the HPRT mutant frequency (Hou et al., 1995, 1999).

I.6. CHROMOSOMAL ABERRATIONS

Analysis of two types of chromosomal aberrations, structural and numerical ones, is carried out by established methodologies. The conventional techniques for chromosomal aberrations, such as G banding, detect both of the aberration types and score the individual subtypes of changes. The chromosome type of change involves both chromatids and is caused by agents such as irradiation in dormant cells (G0 or G1 phase of cell cycle). The chromatid type of damage manifests itself in DNA replication and is the common mechanism for chemicals binding to DNA (Sorsa et al., 1992). Micronuclei are whole chromosomes or their fragments isolated from the main nucleus and thus are relatively easy to score.

The frequencies of structural aberrations, particularly those of dicentric chromosomes, have been extensively used as biological dosimeters of ionizing radiation (Natarajan et al., 1996). Much concerning the stability of chromosomal aberrations has been learnt from exposure to radiation. The lymphocytes carrying dicentic chromosomes are eliminated with a half-life of 150–220 days, as compared to an estimated life span of three years for normal lymphocytes. Balanced translocations are stable and provide a long-term radiation indicator.

Extensive literature exists on carcinogen-induced chromosomal aberrations detected by conventional methods (Sorsa et al., 1992). Biomonitoring studies are many and include, for example, alkenes styrene and butadiene, and the respective epoxides, ethyene and propylene oxide (Sorsa et al., 1992, 1994, 1996; Bonassi et al., 1996; Hayes et al., 2000). In exposure to butadiene, no increase in chromosomal aberrations, micronuclei and sister chromatid exchanges was observed in peripheral lymphocytes, despite the occurrence of butadiene-specific haemoglobin adducts which demonstrated biologically effective exposure (Sorsa et al., 1996; Hayes et al., 2000). In ethylene oxide exposure the most sensitive cytogenetic parameter, sister chromatid exchange, was about one order of magnitude less sensitive than haemoglobin adducts (Tates et al., 1991). Such results suggest that classical cytogenetic end-points may lack the required sensitivity or specificity to respond to low exposures in human biomonitoring studies.

Developments in the field of molecular cytogenetics have made available specific and sensitive tools for the quantification of genetic damage resulting from occupational low level exposures with greater accuracy (Zijno et al., 1994, 1996a,b; Natarajan et al., 1996). The method applicable to environmental populations is based on the analysis of chromosomal rearrangements involving breakage-prone pericentromeric regions, using double hybridization with centro-meric and pericentromeric DNA probes (Eastmond et al., 1994; Rupa et al., 1995). Chromosome-specific fluorescent DNA probes are used in these fluorescent *in situ* hybridization (FISH) techniques, also called chromosome painting techniques. The scoring is easy and unambiguous, and is amenable to complete automation. The methodology has been applied to human biomonitoring of benzene and butadiene exposure (Carere et al., 1995; Hayes et al., 2000). A definite advantage of this experimental approach, in comparison to classical cytogenetic methods, is the possibility to detect chromosomal damage in interphase cells: this enables the scoring of larger cell populations and the analysis of cell types other than white blood cells (Moore et al., 1993). However, the promises of modern cytogenetic techniques in biomonitoring of chemical exposure are yet to come and so far applications are few (Hayes et al., 2000).

The relationship between chromosomal aberrations and cancer has been studied directly in humans who have been assayed for aberrations at any time point and then followed for appearance of cancer (Hagmar et al., 1994; Bonassi et al., 1995; Bonassi et al.,

2000). The subjects who historically had an increased level of chromosomal aberrations were likely to develop cancer more often than those registered with low levels of chromosomal aberrations. Sister chromatid exchange rates were not related to the risk of cancer, while the data for micronuclei were too sparse to allow conclusions.

I.7. SURROGATE VERSUS TARGET TISSUES

In biomonitoring of almost any end-point, surrogate rather than target tissues have to be used. For instance, DNA adducts are usually measured in white blood cells even if the lung is the target. Analogously, protein adducts are measured in haemoglobin, HPRT mutations and chromosomal aberrations in peripheral lymphocytes. The information on the applicability of surrogate tissues in humans is scanty. For white blood cells it is unclear to what extent their own metabolism contribute to the activation of the DNA-binding species. For all blood cells the events in stem cells in the bone marrow, peripheral blood or in extravascular tissue compartments, where lymphocytes reside, are very difficult to distinguish and relate to adducts, mutations and chromosomal aberrations. Even if the target-surrogate tissue discussion below is on DNA adducts, the problem is relevant for other end-points as well. The problem is also very relevant to DNA and RNA adducts excreted in urine, but these points have been discussed in the appropriate section earlier.

In some animal studies the question of the correlation of adducts in blood cells and tissue DNA has been addressed (Eide et al., 1995; Zhao et al., 1999). Rats were exposed by inhalation to individual alkenes from ethene to octene, and DNA adducts in liver and lymphocytes, as well as haemoglobin adducts, were measured (Fig. 7). For all these adducts the levels decreased from ethene to octene. However the decrease was 5-fold in liver, 30-fold in lymphocytes and 2000-fold in haemoglobin. This was interpreted as indicating a complex interplay of tissue uptake, metabolism and diffusion out of organs.

An important factor to be considered in the analysis of surrogate versus target tissues is the solubility of the exposing agent. Thus particle-bound agents, such as PAHs, would be present at high concentrations at the site of contact in the airways as compared to

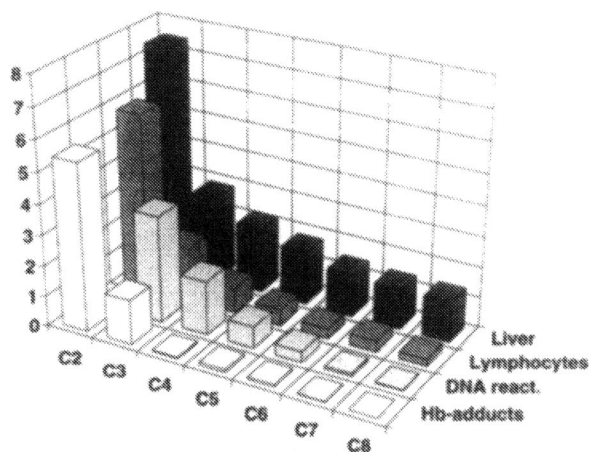

Fig. 7. Liver and lymphocyte DNA adducts (per 10^7 nucleotides) and haemoglobin N-(2-hydroxyalkyl)valine adducts ($\times$ 500 pmol/g) in rats exposed to 300 ppm of alkenes from ethene (C2) to octene (C8) for 12 h on three consecutive days (Eide et al., 1995). (DNA react. = reactivity of alkene epoxides with DNA *in vitro*)

blood and tissue compartments. A soluble agent, such as alkene, would be assumed to be distributed relatively more to all the tissue compartments of the body (Zhao et al., 1999). However, many other properties of the exposing agent, such as lipophilicity and affinity to metabolic pathways, and characteristics of exposure, chronic versus acute, play a role at the level of the DNA and protein adducts. Thus it is important to distinguish at least the particle-bound exposures (caused by PAH-like agents) and the agents soluble in the body fluids (aromatic adducts and soluble adducts, discussed below).

In humans, smoking has been the main model exposure used. Smoking is a known risk factor of laryngeal cancer. Aromatic adducts of laryngeal tissue, obtained from operations, were analysed and there was a relationship to smoking. Both tumour and normal laryngeal tissues showed a correlation of about 0.9 to the total white blood cells (Szyfter et al., 1994).

Tobacco smoke contains methylating and hydroxyethylating principles, originating for instance from tobacco-specific nitrosamines and ethene, respectively. The levels of 7-methylguanine were highest in the bronchial DNA of smokers, exceeding the level in non-smoker almost 4 times (Mustonen et al., 1993). In a small number of smokers both target (bronchial) and surrogate (lymphocyte) DNA were available, showing a correlation of 0.8. Larynx tissue samples obtained from operations have also been assayed for 7-methylguanine DNA adducts. There was a relationship to smoking, and larynx adduct levels exceeded 2-fold those of white blood cells (Szyfter et al., 1996). There was a modest correlation only between 7-alkylguanines and aromatic adducts (discussed above).

Usually biomonitoring studies resort to blood sampling and address the problem of different cell types. Some 25% of white blood cells are long-lived lymphocytes and most of the remainder are short-lived granulocytes. Thus in acute exposure DNA adducts may be present in both cell types but in chronic exposure there may be an increasingly large difference in the adduct level between these cell types. In a study on smokers it was shown that the smokers had 2.5 times more aromatic DNA adduct levels in lymphocytes than non-smokers, whereas in granulocytes no such difference was noted (Savela & Hemminki, 1991). Smokers also had elevated levels of 7-methylguanine in their lymphocytes as compared to the granulocyte DNA, and a similar difference in the cell types were noted for DNA adducts induced by styrene (Mustonen & Hemminki, 1992; Vodicka et al., 1993). In chronic exposures, it is advantageous but logistically difficult to carry out isolation of lymphocytes for DNA adduct analysis. In cases where total blood cell DNA is used, the adducts, nevertheless, probably derive from lymphocytes.

I.8. CONCLUSIONS

Interlaboratory comparisons in methods of DNA and protein adduct determination have shown the need for quality control in any laboratory carrying out these assays. The assays require several steps of preparation and are thus more demanding than standard clinical laboratory assays. The only reasonable way to assure quantitative uniformity of the results is to insert relevant standards in each set of samples analysed. It is more difficult to institute effective quality control to human mutation and chromosomal aberration assays because variation may be introduced in the manipulation of cells before the actual assay in the laboratory. From then on, frozen cells can be used as a reference cell line for the control of day-to-day variation.

There is increasing emphasis on using human biomonitoring data in human health risk assessment. Setting out principles for qualitative cancer risk estimation, the IARC Monographs state in their preamble: "Data relevant to mechanisms of the carcinogenic action are also evaluated... The data may be considered to be especially relevant if they show that the agent in question has caused changes in the exposed humans that are on the causal pathway to carcinogenesis" (IARC, 1996). As DNA and protein adduct studies in humans are becoming more quantitative, they will be useful even for quantitative risk estimation. As epidemiological studies always show the risks of exposure decades previously, adduct studies can be used in present-day risk estimation. The adduct studies are likely to give clues to individual risks and may therefore be useful in protecting sensitive populations. Assay of point mutations, such as those in the HPRT locus have been used to a limited extent in biomonitoring of chemical exposure, and more studies are needed before the usefulness can be established. Mutations, as compared to DNA adducts, are mechanistically closer to the cancer end-point and would provide a valuable addition to risk assessment of chemical exposures. The new cytogenetic methods based on staining of specific chromosomes or subchromosomal regions are evolving and experience will be gained on their applicability in human biomonitoring.

DNA adducts, point mutations and chromosomal aberrations are measurable in humans and are on the causal pathway. Thus, they are highly relevant for cancer risk assessment. However the direct links between these end-points and cancer are limited. As to DNA adducts, a follow-up study has been carried in the Shanghai region, a high incidence area of liver cancer. Urinary aflatoxin-guanine adducts were a clear risk indicator of liver cancer in these individuals (Qian et al., 1994). The relationship between chromosomal aberrations and cancer has been studied directly in humans who have been assayed for aberrations at any time point and then followed for appearance of cancer (Hagmar et al., 1994; Bonassi et al., 1995). The subjects who had historically an increased level of chromosomal aberrations were likely to develop cancer more often than those registered with low levels of chromosomal aberrations.

In the area of apparent endogenous genotoxic damage, biomarkers have provided an indispensable new insight into carcinogenic mechanisms in humans. All the biomarkers used show that even in humans not known to be exposed to external genotoxic agents, specific DNA and protein adducts can be identified. For example, 7-methylguanine and 7-(2-hydroxyethyl)guanine are present in DNA at levels which would require a major environmental exposure (Zhao et al., 1999). How these findings relate to background mutation frequency and chromosomal aberrations will be up to future research to clarify.

I.9. RECOMMENDATIONS AND RESEARCH NEEDS

Considerable resources have been allotted to the technical development and limited human application of exposure and effect biomarkers.

- Establishment by proven techniques of the relative levels of biomarkers in response to exposures would be an urgent task. It is important to relate the induced responses to those found in the background. Inclusion of several biomarkers in a single study would be very helpful but requires sufficient resources and collaboration between scientist.

- Comparisons of biomarker levels to cancer risk in humans and in experimental animals would provide data for risk assessment.

- It would be important to encourage characterization of biomarkers in human populations in terms of kinetics, tissue distribution and background variation. The relationships between target and surrogate tissue need to be established; particle-bound compounds such as PAHs deposited in the lung are important in this respect.

- The relationship of these markers to the cancer end-point needs to be established in follow-up studies, for which blood and tissue banks are very useful.

I.10. REFERENCES

Aarnio M, Sankila R, Pukkala E, Salovaara R, Aaltonen L, De La Chapelle A, Peltomäki P, Mecklin J.-P, & Järvinen H (1999) Cancer risk in mutation carriers of DNA-mismatch-repair genes. Int J Cancer, **81**: 214-218.

Albertini RJ, Nicklas JA, & O'Neill JP (1996) Future research directions for evaluating human genetic and cancer risk from environmental exposure. Environ Health Perspect, **104**(Suppl 3): 503-510.

Alexandrov K, Rojas M, Geneste O, Castegnaro M, Camus A-M, Petruzzelli S, Giuntini C, & Bartsch H (1992) An improved fluorometric assay for dosimetry of benzo(a)pyrene diol-epoxide-DNA adducts in smokers' lung: comparison with total bulky adducts and aryl hydrocarbon hydroxylase activity. Cancer Res, **52**: 6248-6252.

Allam K, Saha M, & Giese RW (1990) Preparation of electrophoretic derivatives of N7-(2-hydroxyethyl)guanine, an ethylene oxide DNA adduct. J Chromatogr, **499**: 571-578.

Bakthavachalam J, Annan R, Beland FA, Vouros P, & Giese P (1990) Selection of electrophoretic derivatives of 1-aminopyrene and 2-aminofluorene for determination by gas chromatography with electron-capture negative-ion mass spectrometry. J Chromatogr, **500**: 373-386.

Bartsch H (1996) DNA adducts in human carcinogenesis: etiological relevance and structure-activity relationship. Mut Res, **340**: 67-79.

Bartsch H, Rojas M, Alexandrov K & Risch A (1998) Impact of adduct determination on the assessment of cancer susceptibility. Recent Results Cancer Res, **154**: 86-96.

Bastlova T, Vodicka P, Peterkova K, Hemminki K, & Lambert B (1995) Styrene oxide-induced HPRT mutations and DNA strand breaks in cultured human T-lymphocytes. Carcinogenesis, **16**: 2357-2362.

Beach A & Gupta R (1992) Human biomonitoring and the ^{32}P-postlabeling assay. Carcinogenesis, **13**: 1053-1074.

Bonassi S, Abbondandolo A, Canurri L, Dal Pra L, De Ferrari M, Degrassi F, Forni A, Lamberti L, Lando C, Padovani P, Sbrana I, Vecchio D, & Puntoni P (1995) Are chromosome aberrations in circulating lymphocytes predictive of future cancer onset in humans? Preliminary results of an Italian cohort study. Cancer Genet Cytogenet, **79**: 133-135.

Bonassi S, Montanaro M, Ceppi M, & Abbondandolo A (1996) Is human exposure to styrene a cause of cytogenetic damage? A re-analysis of the available evidence. Biomarkers, 1: 217-225.

Bonassi S, Hagmar L, Strömberg U, Huici Montagud A, Tinnerberg H, Forni A, Heikkilä P, Wanders S, Wilhardt P, Hansteen I-L, Knudsen LE, & Norppa H (2000) Chromosomal aberrations in lymphocytes predict human cancer independently of exposure to carcinogens. Cancer Res, 60: 1619-1625.

Brauch H, Weirich G, Hornauer MA, Störkel S, Wöhl T & Bruning T (1999) Trichloroethylene exposure and specific somatic mutations in patients with renal cell carcinoma. J Natl Cancer Inst, 91: 854-868.

Bykov V & Hemminki K (1996) Assay of different photoproducts after UVA, B and C irradiation of DNA and human skin explants. Carcinogenesis, 17: 1949-1955.

Bykov V, Kumar R, Försti A, & Hemminki K (1995) Analysis of UV-induced DNA photoproducts by ^{32}P-postlabelling. Carcinogenesis, 16: 113-116.

Bykov VJ, Jansen CT, & Hemminki K (1998a) High levels of dipyrimidine dimers are induced in human skin by solar-simulating UV radiation. Cancer Epidemiol Biomarkers Prev, 7: 199-202.

Bykov VJ, Marcusson JA, & Hemminki K (1998b) Ultraviolet B-induced DNA damage in human skin and its modulation by a sunscreen. Cancer Res, 58: 2961-2964.

Bykov VJ, Sheehan JM, Hemminki K, & Young AR (1999) In situ repair of cyclobutane pyrimidine dimers and 6-4 photoproducts in human skin exposed to solar simulating radiation. J Invest Dermat, 112: 326-331.

Carere A, Antoccia A, Crebelli R, Degrassi F, Fiore M, Iavarone I, Isacchi G, Lagorio S, Leopardi P, Marcon F, Palitti F, Tanzarella C, & Zijno A (1995) Genetic effects of petroleum fuels: cytogenetic monitoring of gasoline station attendants. Mut Res, 332: 17-26.

Cariello NF (1994) Software for the analysis of mutations at the human hprt gene. Mut Res, 312: 173-185.

Castegnaro M & Phillips DH (1997) Interlaboratory standardisation and validation of DNA adduct postlabelling methods for human studies. Final report, Lyon, International Agency for Research on Cancer.

Chaudhary AK, Nokubbo M, Reddy GR, Yeola SN, Morrow JD, Blair IA & Marnett LJ (1994) Detection of endogenous malondialdehyde-deoxyguanosine adducts in human liver. Science, 265: 1580-1582.

Chaudhary AK, Nokubo M, Oglesby TD, Marnett LJ, & Blair IA (1995) Characterization of endogenous DNA adducts by liquid chromatography/ electrospray ionization tandem mass spectrometry. J Mass Spectrom, **30**: 1157-1166.

Clingen PH, Arlett CA, Roza L, Mori T, Nikaido O, & Green MHL (1995) Induction of cyclobutane pyrimidine dimers, pyrimidine(6-4)pyrimidone photoproducts, and Dewar valency isomers by natural sunlight in normal human mononuclear cell. Cancer Res, **55**: 2245-2248.

Cole J & Skopek RT (1994) Somatic mutation frequency, mutation rates and mutational spectra in the human population. Mut Res, **304**: 33-105.

Coller HA, Khrapko K, Torres A, Frampton MW, Utell MJ & Thilly WG (1998) Mutational spectra of a 100-base pair mitochondrial DNA target sequence in bronchial epithelial cells: a comparison of smoking and nonsmoking twins. Cancer Res, **58**: 1268-1277.

Curry J, Karnaoukhova L, Guenette GC, & Glickman BW (1999) Influence of sex, smoking and age on human hprt mutation frequencies and spectra. Genetics, **152**: 1065-1077.

Dallinga JW, Pachen DMFA, Wijnhoven SWP, Breedijk A, van´t Veer L, van Zandwijk N, Maas LM, van Agen E, Kleinjans JCS, & van Schooten F-J (1998) The use of 4-aminobiphenyl hemoglobin adducts and aromatic DNA adducts in lymphocytes of smokers as biomarkers of exposure. Cancer Epidemiol Biomarkers Prev, **7**: 571-577.

Dogliotti E (1996) Mutational spectra: from model systems to cancer-related genes. Carcinogenesis, **17**: 2113-2118.

Dunlop MG, Farrington SM, Carothers AD, Wyllie AH, Sharp L, Burn J, Liu B, Kinzler KW, & Vogelstein B (1997) Cancer risk associated with germline DNA mismatch repair gene mutations. Hum Mol Genet, **6**: 105-110.

Eastmond DA, Rupa DS, & Hasegawa LS (1994) Detection of hyperdiploidy and chromosome breakage in interphase human lymphocytes following exposure to the benzene metabolite hydroquinone using multicolor fluores- cence in situ hybridization with DNA probes. Mutation Res, **322**: 9-20.

Eide I, Hagemann R, Zahlsen K, Tareke E, Törnqvist M, Kumar R, Vodicka P, & Hemminki K (1995) Uptake, distribution, and formation of hemoglobin and DNA adducts after inhalation of C1-C8 1-alkenes (olefins) in the rat. Carcinogenesis, **16**: 1603-1609.

Eide I, Zhao C, Kumar R, Hemminki K, Wu K, & Swenberg J (1999) A comparison between ^{32}P-postlabelling and high resolution GC/MS for determining N7-(2-hydroxyethyl)guanine adducts. Chem Res Toxicol, **12**: 979-984.

Ferreira M, Tas S, dell'Omo M, Goormans G, Buchet J, & Lauwerys R (1994) Determinants of benzo(a)pyrene diolepoxide adducts to albumin in workers exposed to polycyclic aromatic hydrocarbons. Occup Environmen Medicine, **51**: 451-455.

German J & Ellis N (1998) Bloom syndrome. In: Vogelstein B & Kinzzler K (eds), The genetic basis of human cancer. New York, McGraw-Hill, pp 301-315.

Granath FN, Vaca C, Ehrenberg L, & Tornqvist M (1999) Cancer risk estimation of genotoxic chemicals based on target dose and a multiplicative model. Risk Anal, **19**: 309-320.

Greenblatt MS, Bennett WP, Hollstein M, & Harris CC (1994) Mutations in the p53 tumor suppressor gene: clues to cancer etiology and molecular patogenesis. Cancer Res, **54**: 4855-4878.

Groopman J, Hall AJ, Whittle H, Hudson GJ, Wogan GN, Montesano R, & Wild CP (1992a) Molecular dosimetry of aflatoxin-N7-guanine in human urine obtained in the Gambia, West Africa. Cancer Epidemiol Biomarkers Prevention, **1**: 221-227.

Groopman JD, Zhu J, Donahue PR, Pikul A, Zhang L-S, Chen J-S, & Wogan GN (1992b) Molecular dosimetry of urinary aflatoxin DNA adducts in people living in Guangxi autonomous region, People's Republic of China. Cancer Res, **52**: 45-52.

Groopman JD, Wild CP, Hasler J, Junshi C, Wogan GN, & Kensler TW (1993) Molecular epidemiology of aflatoxin exposures: validation of aflatoxin-N7-guanine levels in urine as a biomarker in experimental rat models and humans. Environ Health Perspect, **99**: 107-113.

Grzybowska E, Hemminki K, Szeliga J, & Chorazy M (1993) Seasonal variation of aromatic adducts in human lymphocytes and granulocytes. Carcinogenesis, **14**: 2523-2526.

Hagmar L, Brögger A, Hansteen I-L, Heim S, Hogstedt B, Knudsen L, Lambert B, Linnainmaa K, Mitelman F, Nordenson I, Reuterwall C, Salomaa S, Skerfving S, & Sorsa M (1994) Cancer risk in humans predicted by increased levels of chromosome damage. Cancer Res, **54**: 2919-2922.

Harris C (1996) P53 tumor suppressor gene: at the crossroads of molecular carcinogenesis, molecular epidemiology, and cancer risk assessment. Environ Health Persp, **104**(Suppl 3): 435-439.

Hayes RB, Zhang L, Yin S, Swenberg JA, Xi L, Wiencke J, Bechtold WE, Yao M, Rothman N, Haas R, O'Neill JP, Zhang D, Wiemels J, Dosemeci M, Li G, & Smith MT (2000) Genotoxic markers among butadiene polymer workers in China. Carcinogenesis, **21**: 51-62.

Hemminki K, Grzybowska E, Chorazy M, Twardowska-Saucha K, Sroczynski JW, Putman KL, Randerath K, Phillips DH, Hewer A, Santella RM, Young TL, & Perera FP (1990a) DNA adducts in humans environmentally exposed to aromatic compounds in an industrial area of Poland. Carcinogenesis, **11**: 1229-1231.

Hemminki K, Randerath K, Reddy MV, Putman KL, Santella RM, Perera FP, Young T-L, Phillips DH, Hewer A, & Santella K (1990b) Postlabeling and immunoassay of polycyclic aromatic hydrocarbon-adducts in white blood cells of foundry workers. Scand J Work Environ Health, **16**: 158-162.

Hemminki K, Szyfter K, & Kadlubar FF (1991a) Quantitation of the ^{32}P-postlabeling reaction using N1, N2 and C8 modified deoxyguanosine 3'-monophosphates as substrates. Chem-Biol Interact, **77**: 51-61.

Hemminki K, Szyfter K, Vodicka P, Koivisto P, Mustonen R, & Reunanen A (1991b) Quantitative aspects of 32P-postlabeling. In: Trends in Biological Dosimetry Gledhill BL & Mauro F (eds). New York, John Wiley & Sons, pp 219-228.

Hemminki K, Söderling J, Ericson P, Norbeck HE, & Segerbäck D (1994a) DNA adducts among personnel servicing and loading diesel vechicles. Carcinogenesis, **15**: 767-769.

Hemminki K, Zhang LF, Kruger J, Autrup H, Törnqvist M, & Norbeck HE (1994b) Exposure of bus and taxi drivers to urban air pollutants as measured by DNA and protein adducts. Toxicol Letters, **72**: 171-174.

Hemminki K (1995) DNA adducts in biomonitoring. J Occup Environ Med, **37**: 44-51.

Hemminki K & Vodicka P (1995) Styrene: from characterisation of DNA adducts to application in styrene-exposed lamination workers. Toxicol Letters, **77**: 153-161.

Hemminki K, Autrup H, & Haugen A (1995) DNA and protein adducts. Toxicol, **101**: 41-53.

Hemminki K, Rajaniemi H, Lindahl B, & Moberger B (1996) Tamoxifen-induced DNA adducts in endometrial samples from breast cancer patients. Cancer Res, **56**: 4374-4377.

Hemminki K, Rajaniemi H, Koskinen M, & Hansson J (1997a) Tamoxifen-induced DNA adducts in leucocytes of breast cancer patients. Carcinogenesis, **18**: 9-13.

Hemminki K, Yang K, Rajaniemi H, Tyndyk M, & Likhachev A (1997b) Postlabelling-HPLC analysis of lipophilic DNA adducts from human lung. Biomarkers, **2**: 341-347.

Hemminki K, Dickey C, Karlsson S, Bell D, Hsu Y, Tsai W-Y, Mooney LA, Savela K, & Perera FP (1997c) Aromatic DNA adducts in foundry workers in relation to exposure, life style and CYP1A1 and glutathione transferase M1 genotype. Carcinogenesis, **18**: 345-350.

Hemminki K, Bykov VJ, & Marcusson JA (1999) Re: Sunscreen use and duration of sun exposure: a double-blind, randomized trial. Letter to editor. J Natl Cancer Inst, **91**: 2046.

Herrero-Jimez P, Tomita-Mitchell A, Furth EE, Morgenthaler S, & Thilly WG (2000) Population risk and physiological rate parameters for colon cancer. The union of an explicit model for carcinogenesis with the public health records of the United States. Mut Res, **447**: 73-116.

Hou S-M, Lambert B, & Hemminki K (1995) Relationship between hprt mutant frequency, aromatic DNA adducts and genotypes for GSTM1 and NAT2 in bus maintenance workers. Carcinogenesis, **16**: 1913-1917.

Hou S-M, Yang K, Nyberg F, Hemminki K, Pershagen G, & Lambert B (1999) HPRT mutant frequency and aromatic DNA adduct levels in non-smoking and smoking lung cancer patients and population controls. Carcinogenesis, **20**: 437-444.

Huttner E, Mergner U, Braun R, & Schöneich J (1990) Increased frequency of 6-thioguanine-resistant lymphocytes in peripheral blood of workers employed in cyclophosphamide production. Mut Res, **243**: 101-107.

IARC (1993) Postlabelling methods for detection of DNA adducts. Phillips DH, Castegnaro M, & Bartsch H (eds). Lyon, International Agency for Research on Cancer (IARC Scientific Publication No. 124).

IARC (1994) Scientific Publication No. 125, DNA Adducts: Identification and biological significance. Hemminki K, Dipple A, Shuker DEG, Kadlubar FF, Segerbäck D, & Bartsch H (eds). Lyon, International Agency for Research on Cancer.

IARC (1996) Monographs on the evaluation of carcinogenic risks to humans, vol. 66. Some pharmaceutical drugs. Lyon, International Agency for Research on Cancer.

IARC (1997) Scientific Publication No. 142, Application of biomarkers in cancer epidemiology. Toniolo P, Bofetta P, Shuker DEG, Rothman N, Hulka B, & Pearce N (eds). Lyon, International Agency for Research on Cancer.

Ichiba M, Hagmar L, Rannug A, Högstedt B, Alexandrie A-K, Carstensen U, & Hemminki K (1994) Aromatic DNA adducts, micronuclei and genetic polymorphism for CYP1A1 and GST1 in chimney sweeps. Carcinogenesis, **15**: 1347-1352.

Knudsen LE, Norppa H, Gamborg MO, Nielsen PS, Okkels H, Soll-Johanning H, Raffn E, Järventaus H, & Autrup H (1999) Chromosomal aberrations in humans induced by urban air pollution: influence of DNA repair and polymorphisms of glutathione S-transferase M1 and N-acetyltransferase 2. Cancer Epidemiol Biomarkers Prev, 303-310.

Kumar R & Hemminki K (1996) Separation of 7-methyl and 7-(2-hydroxyethyl)-guanine adducts in human DNA samples using a combination of TLC and HPLC. Carcinogenesis, **17**: 485-492.

Kumar R, Staffas J, Försti A, & Hemminki K (1995) ^{32}P-postlabelling method for the detection of 7-alkylguanine adducts formed by the reaction of different 1,2-alkyl epoxides with DNA. Carcinogenesis, **16**: 483-489.

Kumar R, Vodicka P, Koivisto P, Peltonen K, & Hemminki K (1996) ^{32}P-Postlabelling of diastereomeric 7-alkylguanine adducts of butadiene monoepoxide. Carcinogenesis, **17**: 1297-1303.

Lavrukhin OV & Lloyd RS (1998) Mutagenic replication in a human cell extract of DNAs containing site-specific and stereospecific benzo(a)pyrene-7,8-diol-9,10-epoxide DNA adducts placed on the leading and lagging strands. Cancer Res, **58**: 887-891.

Loeb LA (1994) Microsatellite instability: marker of a mutator phenotype in cancer. Cancer Res, Oct 1, **54**(19): 5059-5063.

Loechler EL (1996) The role of adduct site-specific mutagenesis in understanding how carcinogen-DNA adducts cause mutations: perspective, prospects and problems. Carcinogenesis, **17**: 895-902.

Lunn RM, Langlois RG, Hsieh LL, Thompson CL, & Bell DA (1999) XRCC1 polymorphisms: effects on aflatoxin B1-DNA adducts and glycophorin A variant frequency. Cancer Res, **59**: 2557-2561.

MacKie R (1996) Skin cancer, 2nd edition ed., London, Martin Dunetz.

Malkin D (1998) The Li-Fraumeni syndrome. In: Vogelstein B & Kinzler K (eds.), The genetic basis of human cancer, New York, McGraw-Hill, pp 393-407.

Moore LE, Titenko-Holland N, Quintana PJE, & Smith MT (1993) Novel biomarkers of genetic damage in humans: use of fluorescence in situ hybridization to detect aneuploidy and micronuclei in exfoliated cells. J Toxicol Environ Health, **40**: 349-357.

Mustonen R & Hemminki K (1992) 7-Methylguanine levels in DNA of smokers and non-smokers' total white blood cells, granulocytes and lymphocytes. Carcinogenesis, **13**: 1951-1955.

Mustonen R, Schoket B, & Hemminki K (1993) Smoking-related DNA adducts: ^{32}P-postlabeling analysis of 7-methylguanine in human bronchial and lymphocyte DNA. Carcinogenesis, **14**: 151-154.

Möller L, Grzybowska E, Zeisig M, Cimander B, Hemminki K, & Chorazy M (1996) Seasonal variation of DNA adduct pattern in human lymphocytes analyzed by 32P-HPLC. Carcinogenesis, **17**: 61-66.

Natarajan AT, Boei JJWA, Darroudi F, Van Diemen PCM, Dulout F, Hande MP, & Ramalho AT (1996) Current cytogenetic methods for detecting exposure and effects of mutagens and carcinogens. Environ Health Perspect, **104**(Suppl. 3): 445-448.

Nestmann ER, Bryant DW, & Carr CJ (1996) Toxicological significance of DNA adducts: Summary of discussions with an expert panel. Regul Toxicol Pharmacol, **24**: 9-18.

Neumann H-G (1984) Analysis of hemoglobin as a dose monitor for alkylating and arylating agents. Arch Toxicol, **56**: 1-6.

Nielsen PS, de Pater N, Okkels H, & Autrup H (1996) Environmental air pollution and DNA adducts in Copenhagen bus drivers: Effect of GSTM1 and NAT2 genotypes on adduct levels. Carcinogenesis, **17**: 1021-1027.

Osterman-Golkar S & Bond JA (1996) Biomonitoring of 1,3-butadiene and related compounds. Environ Health Perspect, **104**(Suppl. 5): 907-915.

Osterman-Golkar S, Peltonen K, Anttinen-Klemetti T, Hinso H, Zorcec V, & Sorsa M (1996) Haemoglobin adducts as a biomarker of occupational exposure to 1,3-butadiene, Mutagenesis, **11**: 145-149.

Otteneder M & Lutz WK (1999) Correlation of DNA adduct levels with tumor incidence: carcinogenic potency of DNA adducts. Mutat Res, **424**: 237-247.

Övrebö S, Haugen A, Hemminki K, & Szyfter K (1994) Biological monitoring of exposure to polycyclic aromatic hydrocarbon in an electrode paste plant. J Occup Med, **36**: 303-310.

Övrebö S, Haugen A, Hemminki K, Szyfter K, Drablös PA, & Skogland M (1995) Studies of biomarkers in aluminuium workers occupationally exposed to polycyclic aromatic hydrocarbons. Cancer Detect Prev, **19**: 258-267.

Perera FP, Hemminki K, Grzybowska E, Motykiewicz G, Michalska J, Santella R, Young T-L, Dickey C, Brandt-Rauf P, DeVivo I, Blaner W, Tsai W-Y, & Chorazy M (1992) Molecular damage from environmental pollution in Poland. Nature, **360**: 256-258.

Perera FP, Tang DL, O´Neill JP, Bigbee WL, Albertini RJ, Santella R, Ottman R, Tsai WY, Dickey C, Mooney LA, Savela K, & Hemminki K (1993) HPRT and glycophorin A mutations in foundry workers: relationship to PAH exposure and to PAH-DNA adducts. Carcinogenesis, **14**: 969-973.

Perera FP, Dickey C, Santella R, O'Neill JPO, Albertini RJ, Ottman R, Tsai WY, Mooney LA, Savela K, & Hemminki K (1994) Carcinogen-DNA adducts and gene mutation in foundry workers with low-level exposure to polycyclic aromatic hydrocarbons. Carcinogenesis, **15**: 2905-2910.

Pfeifer GP & Denissenko MF (1998). Formation and repair of DNA lesions in the p53 gene: Relation to cancer mutation. Environ Mol Mutag, **31**: 197-205.

Plna K, Osterman-Golkar S, Nogradi E, & Segerback D (2000) ^{32}P-postlabelling of 7-(3-chloro-hydroxypropyl)guanine in white blood cells of workers occupationally exposed to epichorohydrin. Carcinogenesis, **21**: 275-280.

Poirier MC (1997) DNA adducts as exposure biomarkers and indicators of cancer risk. Environ Health Perspect, **105**(Suppl. 4): 907-912.

Pontén I, Sayer JM, Pilcher AS, Yagi H, Kumar S, Jerina DM, & Dipple A (1999) Sequence context effects on mutational properties of cis-opened benzo[a]phenanthrene diol epoxide-deoxyadenosine adducts in site-specific mutation studies. Biochemistry, **38**: 1144-1152.

Prevost V, Likhachev AJ, Lokionova NA, Bartsch H, Wild CP, Kazanova OI, Arkhipov AI, Gershanovich ML, & Shuker D (1996) DNA base adducts in urine and white blood cells of cancer patients receiving combination chemotherapies which include N-methyl-N-nitrosourea. Biomarkers, 1: 244-251.

Qian G-S, Ross R, Yu M, Yuan JM, Gao YT, Hemderson BE, Wogan GN, & Groopman JD (1994) A follow-up study of urinary markers of aflatoxin exposure and liver cancer risk in Shanghai, People's Republic of China. Cancer Epidemiol Biomarkers Prevention, 3: 519-521.

Rojas M, Alexandrov K, van Schooten F-J, Hillebrand M, Kriek E, & Bartsch H (1994) Validation of a new fluorometric assay for benzo(a)pyrene diolepoxide-DNA adducts in human white blood cell: comparison with ^{32}P-postlabeling and ELISA. Carcinogenesis, 15: 557-560.

Rojas M, Alexandrov K, Auburtin G, Wastiaux-Denamur A, Mayer L, Mahieu B, Sebastian P, & Bartsch H (1995) Anti-benzo(a)pyrene diolepoxide-DNA adduct levels in peripheral mononuclear cells from coke oven workers and the enhancing effect of smoking. Carcinogenesis, 16: 1373-1376.

Ross JA & Nesnow S (1999) Polycyclic aromatic hydrocarbons: correlation between DNA adducts and ras oncogene mutations. Mutat Res, 424: 155-166.

Rupa DS, Hasegawa L, & Eastmond DA (1995) Detection of chromosomal breakage in the 1cen-1q12 region of interphase lymphocytes using multicolor fluorescence in situ hybridization with tandem DNA probes. Cancer Res, 55: 640-645.

Santella R, Hemminki K, Tang D, Paik M, Ottman R, Young TL, Savela K, Vodickova L, Dickey C, Whyatt R, & Perera FP (1993) PAH-DNA adducts in white blood cells and urinary 1-hydroxypyrene in foundry workers. Cancer Epi Biomarkers Prev, 2: 59-62.

Savela K & Hemminki K (1991) DNA adducts in lymphocytes and granulocutes of smokers and nonsmokers detected by the ^{32}P-postlabeling assay. Carcinogenesis, 12: 503-508.

Segerbäck D & Vodicka P (1993) Recoveries of DNA adducts of polycyclic aromatic hydrocarbons in the 32P-postlabelling assay. Carcinogenesis, 14: 2463-2469.

Shibutani S, Suzuki N, & Grollman AP (1998) Mutagenic specificity of (Acetylamino)fluorene-derived DNA adducts in mammalian cells. Biochemistry, 37: 12034-12041.

Shigenaga MK, Aboujaoude EN, Chen, & Ames BN (1994) Assays of oxidative DNA damage biomarkers 8-oxo-2′-deoxyguanosine and 8-oxoguanine in nuclear DNA and biological fluids by high-performance liquid chromatography with electrochemical detection. Methods Enzymol, **234**: 16-33.

Shuker DE & Farmer PB (1992) Relevance of urinary DNA adducts as markers of carcinogen exposure. Chem Res Toxicol, **5**: 450-460.

Somorovska M, Jahnova E, Tulinska J, Zamacnikova M, Sarmanova J, Terenova A, Vodickova L, Liskova A, Vallova B, Soucek P, Hemminki K, Norppa H, Hirvonen A, Tates AD, Fuortes L, Dusinska M, & Vodicka P (1999) Biomonitoring of occupational exposure to styrene in a plastics lamination plant. Mut Res, **428**: 255-269.

Sorsa M, Wilbourn J, & Vainio H (1992) Human cytogenetic damage as a predictor of cancer risk. In: Mechanisms of carcinogensis in risk identification. Vainio H, Magee P, McGregor D, & McMichael AJ (eds.), Lyon, International Agency for Research on Cancer (IARC Scientific Publication No. 116, pp 543-554).

Sorsa M, Autio K, Demopoulos NA, Jarventaus H, Rossner P, Sram RJ, Stephanou G, & Vlachodimitropolulos D (1994) Human cytogenetic biomonitoring of occupational exposure to 1,3-butadiene. Mutat Res, **309**: 321-326.

Sorsa M, Peltonen K, Anderson D, Demopoulos NA, Neumann H-G, & Osterman-Golkar S (1996) Assessment of environmental and occupational exposures to butadiene as a model for risk estimation of petrochemical emissions. Mutagenesis, **11**: 9-17.

Strickland PT, Routeledge MN, & Dipple A (1993) Methodologies for measuring of carcinogen adducts in humans. Cancer Epi Biomarkers Prev, **2**: 607-619.

Szyfter K, Hemminki K, Szyfter W, Szmeja Z, Banaszewski J, & Yang K (1994) Aromatic DNA adducts in larynx biopsies and leucocytes. Carcinogenesis, **15**: 2195-2199.

Szyfter K, Hemminki K, Szyfter W, Szmeja Z, & Banaszewski (1996) Tobacco smoke-associated N7-alkylguanine in DNA of larynx tissue and leucocytes. Carcinogenesis, **17**: 501-506.

Tagessson C, Kallberg M, Klintenberg C, & Starkhammar H (1995) Determination of urinary 8-hydroxydeoxyguanosine by automated coupled-column high performance liquid chromatography: a powerful technique for assaying *in vivo* oxidative DNA damage in cancer patient. Eur J Cancer, **31A**: 934-940.

Tareshima I, Suzuki N, & Shibutani S (1999) Mutagenic potential of α-(N^2-deoxyguanosinyl)tamoxifen lesions, the major DNA adducts detected in endometrial tissues of patients treated with tamoxifen. Cancer Res, **59**: 2091-2095.

Tas S, Buchet J, & Lauwerys R (1994) Determinants of benzo(a)pyrene diolepoxide adducts to albumin in workers exposed to polycyclic aromatic hydrocarbons. Int Arch Occup Environ Health, **66**: 343-348.

Tates AD & Lambert B (eds) (1999) HPRT mutagenesis: aspects of fundamental and applied research. Mut Res Special Issue, 431, no 2.

Tates AD, Grummt T, Törnqvist M, Farmer PB, vanDam FJ, van Mossel H, Schoemaker HM, Osterman-Golkar S, Uebel C, Yang YS, Zwinderman AH, Natarajan AT, & Ehrenberg L (1991) Biological and chemical monitoring of occupational exposure to ethylene oxide. Mut Res, **250**: 483-497.

Tates AD, Grummt T, van Dam FJ, de Zwart F, Kasper FJ, Rothe R, Stirn H, Zwinderman AH, & Natarajan AT (1994) Measurement of frequencies of HPRT mutants, chromosomal aberrations, micronuclei, sister-chromatid exchanges and cells with high frequency of SCEs in styrene/dichloromethane-exposed workers. Mut Res, **313**: 249-262.

Thompson PA, Seyedi F, Lang NP, MacLeod SL, Wogan GN, Anderson KE, Tang YM, Coles B, & Kadlubar FF (1999) Comparison of DNA adduct levels associated with exogenous and endogenous exposures in human pancreas in relation to metabolic genotype. Mutat Res, **424**: 263-274.

Törnqvist M & Ehrenberg L (1994) On cancer risk estimation of urban air pollution. Environ Health Perspect, **102**(Suppl. 4): 173-181.

Törnqvist M & Landin HH (1995) Hemoglobin adducts for *in vivo* dose monitoring and cancer risk estimation. J Occup Environ Medicine, **37**: 1077-1085.

Törnqvist M, Magnusson A-L, Farmer PB, Tang Y-S, Jeffrey AM, Wazneh L, Beulink GDT, van der Waal H, & van Sittert NJ (1992) Ring test for low levels of N-(2-hydroxyethyl)valine in human hemoglobin. Anal Biochem, **203**: 357-360.

van Maanen JMS, Moonen EJC, Maas LM, Kleinjans JCS, & van Schooten FJ (1994) Formation of aromatic DNA adducts in white blood cells in relation to urinary excretion of 1-hydroxypyrene during consumption of grilled meat. Carcinogenesis, **15**: 2263-2268.

van Sittert NJ, Boogaard PJ, Natarajan AT, Tates AD, Ehrenberg LG, & Tornqvist M (2000) Formation of DNA adducts and induction of mutagenic effects in rats following 4 weeks inhalation exposure to ethylene oxide. Mut Res, **447**: 27-48.

Verghis SBM, Essigmann JM, Kadlubar FF, Morningstar ML, & Lasko DD (1997) Specificity of mutagenesis by 4-aminobiphenyl: mutations at G residues in bacteriophage M13 DNA and G→C transversions at a unique dG^{8-ABP} lesion in single-stranded DNA. Carcinogenesis, **18**: 2403-2414.

Vineis P, Malats N, Porta M, & Real FX (1999) Human cancer, carcinogenic exposures and mutation spectra. Mutat Res, **436**: 185-194.

Vodicka P, Vodickova L, & Hemminki K (1993) ^{32}P-postlabelling of DNA adducts of styrene-exposed lamination workers. Carcinogenesis, **14**: 2059-2061.

Vodicka P, Vodickova L, Trejbalova K, Sram RJ, & Hemminki K (1994) Persistence of O6-guanine DNA adducts in styrene-exposed lamination workers determined by ^{32}P-postlabelling. Carcinogenesis, **15**: 1949-1953.

Vodicka P, Bastlova T, Vodickova L, Peterkova K, Lambert B, & Hemminki K (1995) Biomarkers of styrene exposure in lamination workers: levels of O6-guanine DNA adducts, DNA strand brakes and mutant frequencies in the hypoxanthine-guanine phosphoribosyltransferase gene in T-lymphocytes. Carcinogenesis, **16**: 213-216.

Widlak P, Grzybowska E, Hemminki K, Santella R, & Chorazy M (1996) ^{32}P-Postlabelling of bulky human DNA adducts enriched by different methods including immunoaffinity chromatography. Chem-Biol Interactions, **99**: 99-107.

Xu G, Snellman E, Bykov V, Jansen C, & Hemminki K (2000a) Cutaneous malignant melanoma patients have normal repair kinetics of UV-induced DNA damage in skin in situ. J Invest Dermatol, **114**: 628-631.

Xu G, Snellman E, Bykov V, Jansen C, & Hemminki K (2000b) Effect of age on the formation and repair of UV photoproducts in human skin in situ. Mutat Res, **459**: 195-202.

Young AR, Chadwick CA, Harrison GI, Hawk JLM, Nikaido O, & Potten CS (1996) The in situ repair kinetics of epidermal thymine dimers and 6-4 photoproducts in human skin types I and II. J Invest Dermatol, **106**: 1307-1313.

Zhao C, Tyndyk M, Eide I, & Hemminki K (1999) Endogenous and background DNA adducts by methylating and 2- hydroxyethylating agents. Mutat Res, **424**: 117-125.

Zhao C, Vodicka P, Sram RJ, & Hemminki K (2000) Human DNA adducts of 1,3-butadiene, an important environmental carcinogen. Carcinogenesis, **21**: 107-111.

Zijno A, Marcon F, Leopardi P, & Crebelli R (1994) Simultaneous detection of X-chromosome loss and non-disjunction in cytokinesis-blocked human lymphocytes by *in situ* hybridization with a centromeric DNA probe; implications for the human lymphocyte *in vitro* micronucleus assay using cytochalasin B. Mutagenesis, **9**: 225-232.

Zijno A, Leopardi P, Marcon F, & Crebelli R (1996a) Sex chromosome loss and non-disjunction in women: analysis of chromosome segregation in binucleated lymphocytes. Chromosoma, **104**: 461-467.

Zijno A, Marcon F, Leopardi P, & Crebelli R (1996b) Analysis of chromosome segregation in cytokinesis-blocked human lymphocytes: non-disjunction is the prevalent damage resulting from low dose exposure to spindle poisons. Mutagenesis, **11**: 335-340.

BIOMARKERS OF EXPOSURE AND EFFECT FOR NON-CARCINOGENIC END-POINTS

Antonio Mutti

Laboratory of Industrial Toxicology, Department of Clinical Medicine, Nephrology and Health Sciences, University of Parma Medical School, Via Gramsci 14, Parma, Italy

CONTENTS

II.1. INTRODUCTION

Research on biomarkers (also called biological indicators or biological markers) is aimed at overcoming the growing frustration with the limitations of epidemiological studies based on crude estimations of exposure and limited methodology for assessing health outcomes, typically relying on death certificates or insensitive clinical tests. Indeed, in classical epidemiological studies, only such very late effects as clinical disease or death could be investigated. As a result, only etiologic factors occurring with very high relative risks could be depicted, because of misclassifications and inherent biases of observational research. In biomarker research, the major focus of interest has long been on biomarkers of exposure, expected to reduce misclassification deriving from the use of job titles alone or, at best, of point estimates of airborne pollution. More recently, however, earlier more quantitative and sensitive end-points have been sought to identify toxic effects and to address preventive issues. Finally, biomarkers of individual susceptibility are being intensively investigated in both cancer and non-cancer epidemiology, acknowledging that host factors may play a key role in the development and progression of multifactorial diseases.

Despite such encouraging premises, research on biomarkers is facing overwhelming problems, mainly due to inherent ethical issues. Biomarkers of effect and susceptibility represent an intrusion in personal life, possibly resulting in discrimination or stigmatization. Even if such issues were resolved conflict between the worlds of science and preventive medicine would remain. For, whereas science needs hard data on the predictive and prognostic validity of biomarkers, preventive medicine calls for early intervention on the basis of biomarkers envisaged as the forerunners of potentially serious health effects. On the other hand, in order to become useful tools in risk assessment exercises, biomarkers must be validated. Biochemical or molecular epidemiology studies must be as methodologically rigorous as traditional investigations.

In the validation process, several questions must be addressed before biomarker(s) can be extensively used and accepted as scientific tools in quantitative risk assessment: (i) their intrinsic validity, in terms of relevance, stability, sensitivity, specificity, accuracy, and precision; (ii) their mechanistic basis, including application in relevant animal models; (iii) their advantages over other methods, if available, used to characterize exposure and effects; (iv) their interfering factors, which could act as confounders or modifiers; (v) their predictive and prognostic validity in field investigations with proper study design. Perhaps, no single available marker fulfils all of the above criteria. This justifies their prudent use in risk assessment.

II.2   USE OF BIOMARKERS IN RISK ASSESSMENT

Even limiting the fields to occupational and environmental health, a survey on Medline (1991-1996) showed a progression in the number of records mentioning "Risk assessment": 1309 citations were recorded in 1991, whereas as many as 3402 citations appeared in 1997. The key word "Biomarkers" has also showed an increasing trend (from 121 to 454 hits). The combination of "Risk assessment" and "Biomarkers", as compared to "Risk assessment" alone, showed a parallel time course, although it represented just a small fraction or about 1% of the total number of hits mentioning "Risk assessment" (Fig. 8).

Although such a search is probably biased by inaccurate and sometimes inappropriate indexing, it seems that biomarkers are not readily accepted as useful tools for risk assessment, despite their promising features. However, a growing number of publications and books reviewing biomarkers are available (e.g., NRC, 1987, 1989a,b, 1992, 1995; Hulka, 1990; Schulte & Perera, 1993; Travis, 1993, IPCS, 1993; Mutti, 1995) and even new journals devoted to this topic are now regularly published (Biomarkers; Cancer Epidemiology, Biomarkers & Prevention).

II.2.1   Why risk assessment should benefit from the use of biomarkers

Risk assessment is aimed at quantifying the probability that a particular agent or condition will give rise to adverse health effects under specified conditions, depending on: (i) its intrinsic properties that make it a hazard or source of danger; (ii) its use and the corresponding exposure levels; (iii) the number and susceptibility of exposed subjects. Indeed, even very toxic (hazardous) substances

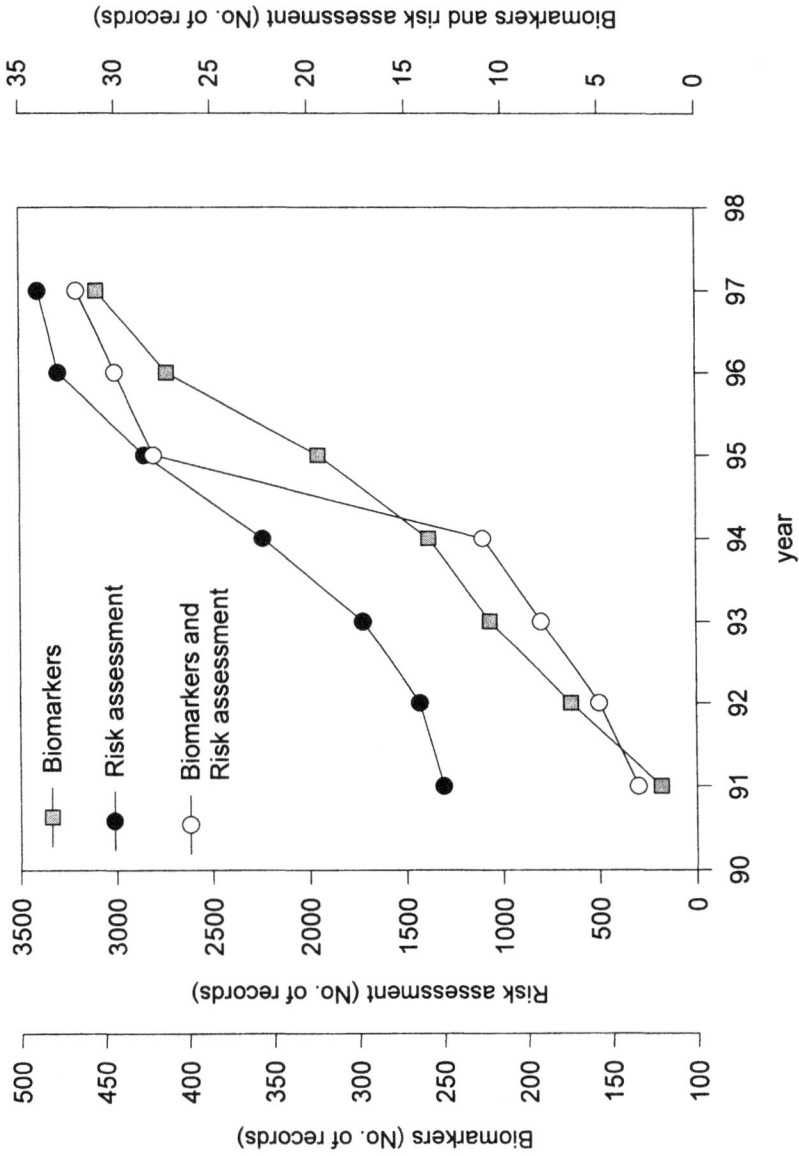

may not pose any significant risk to human health if exposure levels do not exceed a certain threshold for the appearance of adverse health effects or if the number of exposed and susceptible subjects is very low. On the other hand, even substances with low toxicity may cause serious concern when exposure is at levels sufficient to give rise to biologically effective doses and involves a high number of susceptible individuals. The relative importance of each determinant depends on the context. In occupational settings, the number of exposed individuals is usually limited and therefore exposure level is the main determinant. In environmental health, relatively low exposure levels may give rise to a sizeable risk of adverse effects because a large number of susceptible individuals may assume biologically effective doses.

In the framework of risk assessment (hazard identification, dose-response analysis, exposure assessment, and risk characterization) biomonitoring is mainly applied in exposure assessment to identify exposed individuals and groups and to quantify their exposure levels. Biomonitoring is also used in risk characterization to assess the health risks for exposed groups depending on dose levels. Biomarkers of effect can be used in health surveillance programmes for the early diagnosis of exposure-related disease in an individual, but the application of biomarkers of effects or effect monitoring is most often aimed at evaluating whether a well characterized exposure is associated with a shift in the distribution of relevant biochemical or functional end-points indicating early effects on the critical organ or tissue.

II.2.2   Biomarkers and dose-response assessment in humans

A key component of risk assessment is *dose-response assessment,* establishing the probability or degree of response that may be expected from different levels of exposure (NRC, 1987). Acceptable daily intake, environmental guidelines, and occupational exposure limits are most often extrapolated from experimental studies on animals. Uncertainties in extrapolating thresholds across species and dose levels provide a basis for the incorporation of safety factors.

In a deterministic model (i.e., when a theoretical progression exists from exposure, through the absorption and distribution, to early effects and clinical disease), the use of experimental animal data is fully justified. For new chemicals, animal experiments lack suitable alternatives, since risk assessment should obviously precede human exposure. Unfortunately, many adverse effects are multi-factorial in nature. As a result, dose-response relationships for a number of outcomes are inaccurate and experimental research impossible. This is because we do not know or cannot apply risk factors to relevant animal models (Mutti, 1995).

To reduce uncertainties of extrapolation processes, biomarkers may be used to assess dose, effects and susceptibility, thus deriving dose-effect and dose-response relationships in the target species, usually in humans. In these studies, the concentration of the parent compound or its metabolites in accessible biological media or target molecules (biomarkers of internal dose) can be used as independent variables to assess their relationship with biomarkers of (adverse) effect. The latter may be defined not only in terms of clinical manifestations, but also in terms of biochemical and physiological abnormalities measurable long before or well below the appearance of a frank disease (Silbergeld, 1993).

In summary, epidemiological studies rely more and more on biomarkers, that is on indirect measurements of events occurring in a biological system, such as the human body, as a consequence of exposure to environmental pollutants. It is important to bear in mind that biomarkers are surrogate measurements of something difficult or impossible to measure, because it is inaccessible, technically difficult, unacceptably disruptive, or unduly expensive (Mutti, 1993). Although biomarkers are expected to increase the sensitivity of traditional approaches based on crude measures of exposure (e.g., job titles) and of outcome (e.g., death certificates), their validation is a real challenge, due to the intimate nature of surrogate indicators of something difficult or impossible to measure.

For use in preventive medicine, biomarkers should not be regarded as diagnostic tests but rather as indicators that early changes have occurred that could later lead to clinical disease. Those changes must also be completely reversible. To trigger appropriate action, a biomarker finding should be amenable to interpretation with regard to causal factors in the environment, and preventive potential efforts should be realisable, at least in principle (Grandjean, 1995).

101

In this respect, biomarkers are suitable tools to identify ill environments rather than ill people. When dose-effect and dose-response are known, an appropriate biomarker of dose may be sufficient to assess the risk of adverse effects. However, there are situations in which a biomarker of dose cannot be used to predict potential adverse effects. In these situations, biomarkers of effect may be useful to understand whether a shift in their distribution occurred as a consequence of chemical exposure. In this context, biomarkers of effects cannot be used as proof of disease caused by environmental pollution but rather as suitable tools to understand a process that might eventually lead to adverse effects or otherwise unwanted outcomes.

II.3. BIOMARKERS OF EXPOSURE

Biological monitoring and molecular dosimetry are essentially the periodic measurement of a biomarker of exposure to assess the health risk associated with exposure to an industrial chemical. The quantitative nature of the assessment is stressed in the concept of molecular dosimetry, whereas the regular basis of the measurement and the nature of analysed materials are highlighted when the same activity is called biological monitoring.

A biomarker of exposure has been defined as "an exogenous substance or its metabolite or the product of an interaction between a xenobiotic agent and some target molecule or cell that is measured in a compartment within an organism" (NRC, 1989). The term biological indicator of internal dose has essentially the same meaning, but its relationship with adverse effects affecting the critical organ or tissue rather than that with environmental contamination is stressed.

Several biomarkers of exposure may be available for the same chemical. Also, the same biomarker may have different meanings depending on sampling time. The choice should rely on a number of considerations, but mainly on kinetics parameters (Bernard, 1995). For markers with a short half-life (e.g., the concentration of organic solvents in blood), internal dose may mean the amount of chemical absorbed during or shortly before sampling. For markers with intermediate half-life (e.g., urinary metabolites of organic compounds) the internal dose which can be estimated is that occurring during the preceding day(s). For markers with long half-life (e.g., adducts to DNA in lymphocytes or to haemoglobin), internal dose is integrated over a period of months. For cumulative chemicals, internal dose refers to the amount of substance stored in one or more organs and tissues over the years. Sometimes, biomarkers of internal dose reflect the "true" or "effective" dose, i.e., the interaction of reactive metabolites with the critical molecular targets, although such an interaction is measured in non-critical molecules or media. In addition, the same marker may assume different meanings depending on the sampling time.

The most critical issue in the choice of the relevant marker of internal dose and of the relevant sampling time is the knowledge of the mechanistic basis of end-points to be assessed and of the time course of events in the chain connecting exposure with relevant adverse effects. The appropriate biomarker(s) of dose is measurable at a time point when also the outcome of interest can be depicted or attributable to the exposure of interest.

II.3.1 Application of biomarkers to exposure assessment

The use of biomarkers in exposure assessment may be aimed at: (i) unequivocally establishing the fact of exposure in population studies; (ii) reducing misclassification in epidemiological studies; (iii) modelling internal dose, i.e., that occurring at the critical organ, cell or molecule. Whatever the aim, biomarkers of exposure focus on the body burden or on the total dose absorbed, integrating multiple sources of exposure and routes of intake, the pattern of exposure over time, and inter-individual differences in history, habits and behaviours. It has also been noted that biological monitoring is an elegant tool to assess residual exposure when individual protective devices are worn (Droz, 1993). When these factors are not considered to be important, biological monitoring is often questioned, because it may add variability to ambient monitoring. Indeed, biomarkers may vary as a function of time within individuals and at a given time between individuals exposed to the same air concentrations, but kinetics seem to be an important determinant of biological variability relative to exposure patterns. For biomarkers with a half-life of less than 2 h, biomonitoring is not feasible. When the half-life is in the order of 2–10 h, a sample collected at the end of the working day reflects the exposure over the day, while with half-lives of 10–100 h, the optimal sampling time is at the end of the working week, and the results reflect exposure during the preceding few days (Health and Safety Executive, 1992).

For chemicals with long half-lives, most authors agree that biomarkers of exposure provide clear advantages in terms of stability and need a limited number of measurements relative to air levels to characterize exposure (Droz, 1993; Rappaport, 1995). On the other hand, inter-individual and intra-individual variability does not necessarily represent an "annoying detail" in field studies, but rather a central determinant of risk (Hattis, 1996). Therefore, the use of biomarkers and some related practices, such as the expression of urinary concentrations as a function of creatinine, are not primarily aimed at reducing the variance of data. The main aim of biological monitoring is not to reduce, but to explain variance, which is always expected to occur in human populations at risk. Some components of variance, e.g., pre-analytical and analytical factors, must be kept as low as possible, whereas others, such as inter-individual differences in uptake, biotransformation and excretion rate, must be emphasised, the ultimate goal of biomonitoring being the interpretation of data for assessing and managing the associated health risks.

II.3.2   Kinetics and choice of appropriate markers

A broad spectrum of biomarkers of exposure may be available for the same substance, including the concentration of the parent compound or its metabolite(s) in fluids, such as blood, serum and urine, or in other accessible tissues, such as hair and dentine pulp, the adducts of reactive metabolites to DNA, haemoglobin or albumin, and also more sophisticated methods providing indirect estimates of the concentration in critical or storage tissues and organs, such as the kidney cortex for cadmium and the bone for lead.

The choice of the best approach depends very much on the mechanistic basis of adverse effects, which may be classified as: (i) acute or chronic (on the basis of triggering exposure patterns); (ii) local or systemic; (iii) early or delayed (from the triggering exposure); (iv) reversible or irreversible; (v) threshold (dose-related in terms of probability of occurrence and severity) or non-threshold (or stochastic, i.e., depending on the dose for the probability of occurrence but not for its severity). For acute and local effects, biomarkers of exposure may not be actually useful for preventive purposes. Chronic and irreversible adverse effects can result either from cumulated doses or from cumulated effects. Similar cumulated doses may result from repeated short-term, high-dose levels or from long-term, low-level exposures. Although it is often impossible to predict which one of these patterns is relevant to health risks, a relevant biomarker of exposure should serve as a bridge linking occupational or environmental pollution with a long-term health outcome or with some relevant intermediate end-point (Mutti, 1995). If the mechanism of toxicity is known, kinetic parameters are useful to identify the biomarkers suitable for assessing exposure levels that are thought to elicit the observed effects. It is generally assumed that the longer the half-life of a marker, the better is its correlation with most situations representing a matter for concern in public health, i.e., with effects resulting from chronic, long-term, low-level exposure to cumulative toxicants.

A list of biomarkers of internal dose distinguished on the basis of kinetic parameters is shown in Table 9 (adapted with modifications from Bernard, 1995). The half-life is the main quantitative parameter derived from mathematical models to summarize the behaviour of xenobiotics in biological systems. It can be derived from: (i) empirical mathematical formulae fitting experimental data; (ii) compartment models incorporating rate constants derived from experimental data; (iii) simulation models in which variables and constants have a physiological and metabolic basis (Droz, 1986). Sophisticated models based on physiological and pharmacokinetic parameters incorporate a number of host factors accounting for both inter- and intra-individual variability of results (e.g., body build, physical work load, liver and renal function). On the basis of their biological half-life, biomarkers of exposure can be classified into four categories. Such a list is far from being complete, but it is useful to understand the possible use of individual biomarkers.

Information on some metals is not included because of the complexity of metabolic patterns depending on particle size, solubility and oxidation state of inhaled particles. Indeed, element speciation, defined as the capability of separating, identifying and determining the species in which an element is present or (bio)transformed, represents a real challenge for the future of biomonitoring. There are great differences between different species

Table 9. Half-lives of different biomarkers of exposure[a]

	Half-life	Chemical	Indicator	Sample
Very short	2.5 h	benzene	benzene	blood & exhaled air
	3.5 h		phenol	urine
	5 h	carbon monoxide	carboxyhaemoglobin	blood
Short	14 h	n-hexane	2,5-hexanedione	urine
	5 h	styrene	mandelic acid	urine
	19 h		phenylglyoxylic acid	urine
	96 h	tetrachloroethylene	tetrachloroethylene	blood & exhaled air
	96 h		tetrachloroacetic acid	urine
	18 h	polycyclic hydro-carbons	pyrenol	urine
Long	30 days	lead	lead	blood
	18 days	mercury	mercury	blood & urine
	100 days	cadmium	cadmium	blood
		electrophilic inter-	adducts to:	
	120 days	mediates (excluding	haemoglobin	blood
	180 days	repair and other	mononuclear cell DNA	blood
		interfering mecha-		
	20 days	nisms, e.g., steady state)	albumin	serum
Very long	> 10 years	cadmium	cadmium	urine
	> 10 years		cadmium	kidney cortex
	5 years	lead	lead	bones
	2 years	hexachlorobenzene	hexachlorobenzene	serum

[a] Adapted with modifications from Bernard (1995)

of the same metal regarding not only its local and systemic effects, but also its biotransformation. For some metals, e.g., arsenic, specific indicators are available to assess exposure to organic and inorganic compounds, respectively (Lauwerys & Hoet, 1993). However, it has recently been noted that whereas in most instances sensitive and specific analytical techniques are now available to deal with speciation, the main difficulty remains to understand the biological significance of the different species in the toxic response (Apostoli, 1997).

The importance of solubility and metal speciation has been highlighted in biomonitoring studies in workers exposed to cobalt compounds. The urinary excretion of cobalt at the end of the working week is considered to be a good reflection of recent exposure to this element when exposure is to cobalt metal, soluble cobalt salts or hard metals, but not in the case of cobalt oxide which is much less soluble in biological media and persists longer in the lung compartment (Lison et al., 1994).

Another difficult example where speciation and oxido-redox processes have a significant impact for the biomonitoring is exposure to chromium compounds, which is usually assessed by the measurement of the urinary concentration of the element. However, one must acknowledge that the value of this biomarker is rather limited because, even when sampling and analytical problems can be adequately controlled, it poorly reflects exposure of the target tissue to the species of greater concern, i.e., Cr(VI), and gives little information on health risk.

The chemical species is also an important determinant of the biological half-life. Whereas most chromium(III) compounds show a long half-life (> six months), certain soluble compounds of chromium(VI) clear rapidly. However, detailed kinetic studies of urinary chromium suggest that even for soluble chromium(VI) compounds the elimination curve is consistent with the existence of at least three compartments showing half-lives of 7 h, 15–30 days and 3–5 years (Welinder et al., 1983; Rahkonen et al., 1983; Kalliomäki et al., 1985; Aitio et al., 1988). Depending on the sampling time and strategy, urinary chromium may reflect either past or recent exposure (Mutti et al., 1984c). Recent work performed in the USA shows that, contrary to what is usually stated in textbooks of toxicology, the diffusion of Cr(III) through biological membranes is not negligible and it can relatively easily diffuse through the red blood cell (RBC) membrane. In contrast, since Cr(VI) taken up by the RBC is reduced to react with Hb, it cannot be further exchanged with other compartments of the body. Therefore, in case of exposure to Cr(VI) one could expect that the RBC content will remain elevated at least during 3–4 months (the lifetime of a RBC), whereas when exposure is to Cr(III), the RBC concentration of the element will rapidly decrease after cessation of exposure (Finley et al., 1997).

Macromolecule adducts originating from electrophilic metabolites of genotoxic compounds are very useful, because they reflect the dose escaping detoxification and reaching the target protein or DNA. Historically, methods to measure adducts formed by electrophilic intermediates were developed to determine the target or the "true" dose of genotoxic compounds (Ehrenberg & Osterman-Golkar, 1980). As RBCs have a long life (approximately 4 months in humans) binding to haemoglobin is considered to be a good biomarker to measure cumulative internal dose due to repeated exposures and hence to assess the risk also of non-cancer end-points (Costa, 1996). Albumin adducts have a shorter half-time in blood (approximately 20 days) and, therefore, they reflect a more limited period of exposure. However, one advantage of albumin over DNA adducts is that potentially active metabolites interact with this protein upon their release in the blood stream without having to penetrate a cell membrane or to be biotransformed in the cell used for monitoring purposes (Henderson et al., 1989). On the other hand, if adducts are meant to reflect the cellular level at the target site, then haemoglobin adducts would be a more accurate biomarker of target tissue dose (Costa, 1996).

Since their formation and repair rate is probably genetically determined, adducts may also serve as biomarkers of susceptibility. Although only in principle, macromolecule adducts are close to the ideal biomarker sought for risk assessment: they provide an integrated measure of exposure, they represent early and usually reversible biomarkers of effect, and they are likely to reflect individual susceptibility to etiological agents, ultimately to provide an estimate of the individual risk. Indeed, an ideal biomarker of exposure should be "specific for a chemical, detectable in small quantities, measured by non-invasive techniques, inexpensive, associated with prior exposure and which has an excellent positive predictive value to a specific disease status" (Henderson et al., 1989). In practice, adduct measurements are time-consuming, difficult to perform and to standardize, and are limited to compounds or metabolites forming covalent bonds with nucleophilic macromolecules. Also, as in the case of urinary metallothionein, an inducible low molecular weight protein reflecting both Cd body burden and tubular dysfunction (Shaikh & Tohyama, 1984), such "universal" biomarkers do not allow a clear distinction between exposure and effect, ultimately hampering the definition of dose-effect/response relationships and risk assessment.

In summary, knowledge of the mechanism(s) eventually leading to ultimate toxic end-points should drive the choice of the relevant biomarker of exposure. A suitable biomarker of exposure should be correlated with exposure to a single chemical. However, if exposure can be measured, the need for biomonitoring has limited scope. Biomarkers of dose may be more useful as independent variables to assess dose-effect and dose-response relationships. Indeed, "the predictive value of a biomarker of dose relies on its ability to predict adverse effects rather than on its ability to predict exposure" (Mutti et al., 1996).

II.3.3 Validity of biomarkers of exposure

Validity has been defined as the degree to which the results of a measurement correspond to the true state of the phenomenon being measured. Another word for validity is accuracy. However, accuracy as well as other attributes often considered, such as sensitivity and specificity, may have different meanings depending on the scope. These terms – accuracy, sensitivity and specificity – may refer to inherent properties of the biomarker, i.e., to the accuracy with which a biomarker reflects the true exposure level, to its sensitivity in terms of changes associated with varying exposure levels, and to its specificity in terms of selectivity or source specificity. The same terms – accuracy, sensitivity and specificity – may refer to the validity of analytical methods, whereby accuracy is the ability to identify the true value of an accepted standard, sensitivity stands for the dynamic range of the method, and specificity for its ability to measure exclusively the chemical (parent or metabolite) of interest. Finally, accuracy, sensitivity and specificity may refer to the context of application, with regard to the ability of the biomarker to predict adverse effects, and to the number of false negative and false positive individuals in field applications.

II.3.3.1 Intrinsic validity

Relevance and stability are perhaps the most important properties making a biomarker of dose suitable for field applications.

In this respect, the concentration of the parent compound in biological media is generally preferable to that of metabolites which can be shared with other substances. Provided that sampling strategies and storage procedures are carefully planned, the parent compound is usually better correlated with exposure as compared to its metabolites (Ikeda, 1999). However, the parent compound tends to have a shorter half-life, is often volatile, and is usually unrelated to adverse effects, which occur as a consequence of its biotransformation and metabolic activation. If the focus is on dose-response relationships, urinary metabolites have successfully been used as suitable biomarkers of dose (Mutti et al., 1984a).

II.3.3.2   Analytical validity

Analytical validity refers to the properties of analytical methods, including their selectivity, dynamic range, inaccuracy and imprecision. Although simple colorimetric methods may still be used in a few instances, there is a tendency to rely on more sophisticated and costly techniques. Technology and powerful software applied to mass spectrometry are greatly improving hyphenated techniques, all relying on mass spectrometry (MS) detection. Hyphenated techniques take advantage of prior separation of analytes, but especially of: (i) the selectivity and sensitivity of MS; (ii) a wider dynamic range compared to suitable alternatives; (iii) little sample manipulation, which could give rise to artefacts. Whereas nanomolar concentrations can now be measured in biological media, new problems may arise in the control of pre-analytical factors and in the interpretation of the toxicological meaning of such low levels.

Simpler and well validated analytical techniques can however reach sufficient sensitivity and specificity for routine monitoring in occupational settings and, in some cases, even in the environmental range relying on gas or liquid chromatographic separation combined with relatively specific quantification techniques such as photo-ionization, electron capture, and electrochemical detectors. In the case of metals, inductively coupled plasma – mass spectrometry is a promising technique, especially in element speciation, although flameless atomic absorption spectrometry still remains valid for the analysis of trace elements in biological media, especially if discrimination between different compounds of the same element is not required.

II.3.3.3 *Context validity*

In identification of exposure, the chemical itself is certainly more specific than any of its metabolites, which may derive from multiple sources, sometimes including confounders. In most instances, analytical methods are available to quantify both organic and inorganic xenobiotics in the nanomolar range. At such levels, confounding is likely to occur if the possible contribution from other sources, e.g., smoking habits, is not ruled out. Also, local reference values must be defined and updated, since large geographical and temporal variation are known to occur as a consequence of changing patterns of environmental pollution.

For quantitative exposure assessment, consideration should be given to other aspects, including kinetics, which may influence the ability of the biomarker to reflect exposure occurring during the working day. An example is styrene, whose metabolites mandelic and phenylglyoxylic acids have consistently been shown to mirror daily exposure, despite their limited specificity (both ethylbenzene and styrene oxide share the same end-products of biotransformation). Styrene exposure may be estimated with confidence on the basis of these metabolites (Aitio & Kallio, 1999), a large amount of consistent data being available in the literature.

For risk characterization, toxicodynamic factors and especially available dose-response relationships must be taken into account. From this point of view, we must recognise that there are just a few traditional and widely accepted examples of biomarkers of dose that can be used to predict adverse effects, e.g., blood lead and urinary cadmium, although attempts to apply the same methodology (assessment of dose-response relationships based on biomarkers) have been made and will be illustrated at the end of this chapter.

II.3.4 Validation

Validation is an empirical process accumulating evidence that a biomarker of exposure can be used either to assess past exposure or to predict expected outcomes. Such an empirical process encompasses several tiers, from the laboratory development to the field application. During a recent International Symposium on Biomarkers, a workshop was convened to assess the criteria and quality requirements for valid biomarker measurement. Quality assurance has been identified as a key issue or a pre-requisite for the validation of a biomarker of exposure. Quality assurance has been defined as the overall measures taken to ensure that laboratory results are reliable, including the adoption of scientifically sound criteria in the selection of the appropriate biomarker, and technically sound practices not only in the collection, transport, storage and analysis but also in the recording, reporting and interpretation of results. Quality control, either internal or preferably external, is an essential part of quality assurance, in that it aims at verifying that analytical results issued by the laboratory meet the requirements of the user (Aitio & Apostoli, 1995).

Subsequent steps in the validation process include the critical assessment of different relationships: exposure-dose, dose-response (effect), effect-disease. Each one of such relationships may include the possible role of susceptibility biomarkers as potential confounders or modifiers. Whereas for short-lived organic compounds there are many examples of validation studies dealing with exposure-dose and only a few pertaining to dose-response relationships, the opposite is true for relatively long-lived metals. This means that data are often available for information that is easy to assess but unnecessary, whereas it is missing for crucial information, which may be difficult to obtain. Also, whereas in metal toxicology it is widely acknowledged that biomarkers of dose should not necessarily relate to recent exposure, but rather should predict adverse effects, biomonitoring of volatile organic compounds has mainly been used to demonstrate that internal dose reflects exposure. Since the latter can be easily measured, the role of biomarkers as a surrogate measure of something difficult or impossible to measure is questionable. Also, there are a number of situations where skin is a major route of absorption. Moreover, the use of personal protective devices at the workplace is growing and in such cases ambient exposure–dose relationships are either difficult to assess or irrelevant.

II.4. BIOMARKERS OF EFFECT

A biomarker of effect has been defined as "a measurable biochemical, physiological or other alteration within an organism that, depending on magnitude, can be recognised as an established or potential health impairment or disease" (NRC, 1989). Research on biomarkers of effect is rapidly generating a large amount of data measuring intermediate end-points occurring probably after exposure and possibly before illness (Mutti, 1995). Such biomarkers are expected to reflect early modifications preceding progressive structural or functional damage at the molecular, cellular and tissue level. Therefore, they should identify early and reversible biochemical events that may also be predictive of later response (Mutti, 1991; Silbergeld, 1993). Unfortunately, the mechanism of action of many chemicals is unknown at present. Furthermore, changes occurring in target tissues or cells may not be mirrored by biochemical changes occurring in peripheral, accessible media. Finally, whereas early damage may be repaired and subsequent dysfunction compensated for, it may also trigger a "cascade of events" eventually leading to clinical disease (IPCS, 1991). However, there is no magic marker that can be used to distinguish reversible from progressive damage, but rather patterns that need to be identified on a one-by-one basis for each chemical and system or target organ (Mercier & Robinson, 1993).

Three main strategies have been followed in developing biomarkers of effect: (i) epidemiological; (ii) clinical; (iii) experimental. Epidemiological studies have been useful to identify biomarkers that are associated, though not necessarily causally related, to later outcomes. Such a strategy is possible when the outcome is relatively frequent and multifactorial in nature, and the measured biomarker is inexpensive and readily available, e.g., serum cholesterol in cardiovascular disease. Most biomarkers of effect have been identified on the basis of pathophysiological reasoning, usually starting from clinical conditions, extrapolating backward changes preceding illness, e.g., early markers of nephrotoxicity. Since such markers are then used in epidemiological studies, a different methodological context of application may lead to misinterpretation of the health significance of observed changes, which greatly depends on the prevalence of the condition being examined (Mutti, 1993). The experimental approach is usually multi-tiered: *in vitro* studies and animal experiments are used to identify the mechanism of action of toxic chemicals; comparative studies are performed to verify that changes in candidate biomarkers occur both in target tissue and in peripheral, accessible media; epidemiological investigations are carried out to assess the sensitivity of such potential biomarkers to toxic chemicals. Such an approach appears to be the most relevant to immuno- and neurotoxicology, for in the immune and in the nervous system, potential target cells are either disseminated in the organism or confined to an inaccessible compartment. In both cases, function is regulated by the balance between triggering and inhibiting stimuli. Any alteration of such systems can hardly be interpreted in terms of toxicity in the absence of experimental models providing some mechanistic clues.

The non-identity of correlation and causation is often forgotten and sometimes disregarded in the selection of relevant biomarkers. Selection of an inappropriate or pseudo-biomarker can have consequences ranging from wasted time and resources to the arrival at false conclusions and even erroneous decisions in public health policy, e.g., the setting of inappropriate reference and guideline values, mistaken priority setting, and inadequate or excessive allocation of resources (Nolan, 1995). Because of the considerable social and economic impact, several work- and environment-related non-cancer end-points have been identified as priority issues in Public Health and have been the focus of biomarker research aimed at developing suitable tools and information for use in risk assessment. In the following sections, a summary of biomarkers available to assess nephrotoxicity, neurotoxicity, and pneumotoxicity will be given. For reproductive toxicity and immunotoxicity some encouraging results have been obtained, but much work remains to be done because of inherent difficulties in the search for relevant biomarkers and in validation studies (NRC, 1989a, 1992; Savitz & Harlow, 1991; Nolan, 1995).

II.4.1 Nephrotoxicity

Work on biomarkers of nephrotoxicity dates back to the mid-twentieth century, when Friberg's pioneering studies on cadmium nephrotoxicity lead to the set-up of a qualitative test identifying low molecular weight proteinuria (Friberg, 1950). It took 15 years to develop semi-quantitative methods to assess cadmium-induced low-molecular weight proteinuria (Piscator, 1962, 1966a,b) and 15 years more to characterize cadmium-induced proteinuria on the basis of the urinary excretion of single low and high molecular weight serum proteins and enzymes (Bernard et al., 1979). Immunochemical methods available by the early 1980s lead to the identification of kidney-derived antigens as early markers indicating that tubular cell damage and not simply dysfunction was associated with chronic exposure to cadmium (Mutti et al., 1984a, 1985). We are just at the beginning of a new era, in which basic science is dominated by molecular biology, and new technologies requiring powerful software are being developed, e.g., LC-MS. It is likely that new diagnostic tools will become available, after some years of latency necessary to transfer basic knowledge into applied research. However, information gained thus far can be used to draw some provisional conclusions.

Biomarkers of nephrotoxicity have been reviewed in several workshops (e.g., CEC-IPCS, 1989; Mueller et al., 1997) and books (e.g., CEC-IPCS, 1989; IPCS, 1991; NRC, 1995). As in other areas of biomarker research, the main conclusion of such reviews is that broad batteries of tests have been developed, including very sensitive end-points and site-specific biomarkers useful to characterize toxic damage at the tissue, cellular and molecular levels. As for other situations, e.g., for neurobehavioural tests, a recent USA/EU joint workshop was aimed at standardizing and sharing methods, and possibly at identifying a core test battery suitable for use in population studies (Mueller et al., 1997). After extensive discussion, the need for better and more versatile biomarkers was stressed, particularly for extracellular matrix markers indicating a potential progression of early damage towards long-term outcomes. A core battery of urinary markers, among those commercially available, has been recommended, including albumin, one low-molecular weight protein, such as β2-microglobulin or RBP, and one marker of cytolysis, such as the activity of the lysosomal enzyme NAG (*N*-acetyl-β-D-glucosaminidase). The meaning of such markers, together with ambiguities when observed changes are small, is summarized in Table 10. The interval roughly corresponding to the 95th-99th percentile of reference values is also an area of overlap between "normal" and "pathological" values. The small deviations from reference values falling within this interval cannot be interpreted at the individual level, since alternative explanations are possible. When such markers are examined on a group basis, in the context of epidemiological studies, potential confounding by factors mentioned in the last column should be excluded.

More detailed studies require a much larger battery including confirmatory tests and additional markers to identify site-specific injury (Table 11).

Table 10. Meaning of renal markers depending on deviation
from reference values

Marker	Meaning	Range of values or percentile of the reference distribution[a]	Alternative meaning
Albuminuria	glomerular lesions implying increased glomerular permeability	20-200 mg/g creatinine	Impaired tubular reabsorption; interference by a number of factors (workload, meat meal, etc.)
		>200 mg/g creatinine	none
Low molecular weight proteinuria (β2-microglobulin, RBP)	tubular lesions giving rise to reduced tubular reabsorption	300-1000 µg/g creatinine	overload proteinuria; competition for tubular uptake (meat meal, etc.)
		>1000 µg/g creatinine	none
NAG	leakage from damaged tubular cells	>95th-99th	exocytosis; interference with enzyme activity by inhibitors
		>99th	probably none

[a] Reference values may vary depending on the laboratory

These markers include brush-border antigens from the convoluted part of the proximal tubules, intestinal alkaline phosphatase from the straight part of proximal tubules, fibronectin and laminin fragments from the interstitial matrix. Additional markers, such as prostanoids, growth factors and cytokines need further experimental and clinical validation.

II.4.1.1 Validation of renal biomarkers

Full validation of biomarkers should rely on follow-up studies indicating the health significance of observed changes. Micro-albuminuria and low-molecular weight proteinuria fulfil this condition in subjects suffering from diabetes mellitus and from chronic cadmium poisoning, respectively. In such situations, both markers are predictors of an accelerated deterioration of renal

Table 11. Site-selectivity of biomarkers of early renal effects

- *Markers of effects at the glomerular level*:
    - High-molecular weight plasmaproteins
      (molecular weight > 40 000) in urine:
        - Albumin; Transferrin; IgG
    - Components of glomerular structures in urine/plasma:
        - Fibronectin; Laminin; Sialic acid
        - Prostanoids: TXB_2, 6-keto-$PGF_{1\alpha}$
- *Markers of early effects at the proximal tubule*:
    - Low-molecular weight proteins in urine:
        - β2-microglobulin; RBP; α1-microglobulin; CC16 (Protein 1)
    - Enzymes: NAG, TNAP, IAP, γGT, AAP
    - Kidney antigens
        - Brush-border: BB-50, BBA, HF5
        - Proximal tubule: IAP, GST α
        - Distal tubule: GST π
- *Marker of the distal tubule: kallikrein in urine*
- *Marker of the loop of Henle: Thamm-Horsfall glycoprotein*
- *Markers of the collecting tubule and interstitium: $PGF_{2\alpha}$ and PGE_2*
- *Site-unrelated markers: GAGs*

function. Outside of these two situations, no information is available to interpret the early changes resulting from chemical exposure and there is a need to conduct longitudinal studies on populations with well-characterized exposure or risk. However, it has been noted that, perhaps with the exception of chemicals with a very long half-life, the long-term significance of early changes is very difficult to assess (Mutti, 1995). Therefore, when persistent microproteinuria is observed in the context of a documented chronic exposure to a suspected or established nephrotoxicant, it is prudent to consider that it might have a similar meaning as in incipient diabetic or cadmium nephropathy (Bernard et al., 1979).

Such a view is corroborated by recent animal experiments, which support epidemiological studies suggesting that hydrocarbon exposure can accelerate the progression of renal disease towards chronic renal failure (Mutti et al., 1999). In these experiments, the urinary excretion rate of albumin and fibronectin, but to some extent also of low-molecular weight proteins, correlates with the histopathological semiquantitative scoring for interstitial infiltration and fibrosis (Fig. 9).

Although these findings cannot be generalized, there is growing evidence indicating that chronic hyperfiltration of proteins and the subsequent overload to proximal tubules may stimulate tubular cells to release factors initiating the cascade eventually leading to interstitial fibrosis and chronic renal failure (Abbate et al., 1998). This mechanism would be involved in several human and experimental renal diseases where proteinuria is the main factor accelerating progressive renal disease (for a review, see Remuzzi & Bertani, 1998).

II.4.2 Neurotoxicity

Concern for the acute and long-term effects of chemicals on the nervous system has been growing because of: (i) the vulnerability of the central and peripheral nervous system to a broad spectrum of chemical pollutants; (ii) the inability of neurones to regenerate after injury, which accounts for the serious consequences of disabling conditions caused by neurotoxicity; (iii) the unique and critical role of the nervous system in the control of body function, including that of other organs and systems, such as the endocrine and the immune system, which suggests that consequences of neurotoxicity might go far beyond neuronal damage and impairment. Anger & Johnson (1983) identified 750 chemicals for which, at some concentration, there is evidence of adverse effects measured by established tests of nervous system chemistry, structure or function.

A number of reviews discussing behavioural and biochemical markers of neurotoxicity have been published in recent years (e.g., Anger, 1990, 1993; Silbergeld, 1993; Costa, 1996; Manzo et al., 1996). Studies combining several approaches including neurophysiological, behavioural, and neurochemical assays are likely to provide the integrated information necessary to assess health risks.

A review of all available behavioural tests is out of the scope of this text. This does not mean that behavioural tests are not objective and reliable, but simply that recent biomarker research in this area focusses on markers providing insights on the mechanistic basis of selective chemical injury to critical or selectively vulnerable target

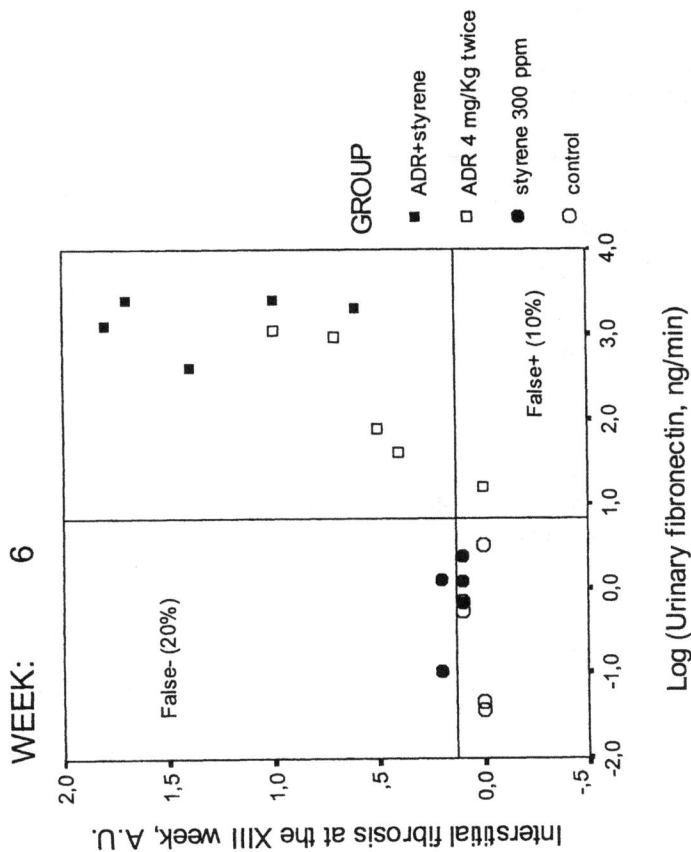

Fig. 9. Relationship between the urinary excretion of fibronectin at the 6th week and the histopathological scoring for interstitial fibrosis at the end of the experiment

cells and systems. Whereas behavioural test batteries are species-specific, measurements over time of neurochemical markers may be helpful in animal studies to assess selected non-conventional end-points such as early manifestations of late-developing toxic effects, recovery processes or adaptive changes during chronic exposure. Quantitative information provided by these data could be valuable for extrapolation across species and dose levels, ultimately for risk assessment. For example, the process of inferring potential hazards to humans from animal studies might be significantly aided by demonstrating parallel biomarker patterns, whereas lack of concordant biomarker patterns may indicate the mediation of dissimilar underlying processes and biological end-points.

In humans, biochemical markers of neurotoxicity may offer an important complement to the existing clinical and laboratory methods for neurological examination in cases of severe poisoning. More complex is their use in assessing subtle changes such as those occurring in chronic low-level exposure to environmental chemicals. Since detection of neurotoxicity depends in most instances on recognising patterns of changes rather than any single abnormality, it seems unlikely that any one laboratory alteration will correlate precisely with the clinical or subclinical entities of disease (Manzo et al., 1996). In general, the earlier the marker in the progression of biological response, the less strongly it can be expected to predict later outcome but the more useful it may be for purposes of prevention. Table 12 shows some biomarkers used in field studies so far. In the future, the availability of a wider array of reliable markers might permit the dissection of complex biological processes into a biologically based dose-response paradigm. In this context, the concept of assessing neurotoxicity may become more focused even in the case of chronic low-level exposures. This should help to establish which markers offer the greatest promise for application in biomonitoring programmes for chemicals representing major occupational or public health hazards.

At present, use of neurochemical processes as a source of biomarkers is limited to agents whose mode of action is sufficiently understood. Therefore, prospects for significant advances in this field are linked to mechanistic studies, in particular to identification of molecular targets that are present not only in the nervous tissue but also in easily accessible sites such as plasma and peripheral blood cells.

Table 12. Peripheral blood parameters used as surrogate markers
of nervous system function

Hormones

prolactin

Autoantibodies

to myelin basic protein

Receptors

lymphocyte muscarinic acetylcholine receptors,

β-adrenoceptors, sigma receptors; platelet α2-adrenoceptors

Enzymes

red cell cholinesterase, lymphocyte neurotoxicity target
esterase (NTE), platelet type-B monoamine oxidase (MAO-B)

Uptake systems

5-HT uptake in platelets

Signal transduction components

intracellular free calcium ion concentration,
adenylyl cyclase, phosphoinositide metabolism in platelets and
lymphocytes

This conclusion is supported by a number of experimental data:
(i) Processes such as neurotransmission or cell signalling that are
involved in a variety of toxic events in the brain are ubiquitous and
can be examined not only in the nervous system, but also in other
tissues, including platelets and peripheral blood lymphocytes. (ii)
There is increasing evidence that neurobehavioural and endocrine
processes are closely correlated, and recent discoveries add a new
dimension to our understanding of the neuroendocrine environment
in which certain subtle responses to drugs and chemicals take place.
These links open up opportunities for investigating the nervous
system indirectly by means of minimally invasive procedures. (iii)
Although the nervous system is shielded from the blood by
anatomical barriers (e.g., blood-brain and blood-nerve barriers), a
breakdown of these barriers causing "leakage" of neural macromolecules and immune cells may occur during nervous system
injury. Quantitative assessment of such circulating indicators in
biological fluids might be a valuable tool not only in experimental
toxicology but also in clinical practice to identify cell-specific
pathological changes or to follow-up patients with overt neurotoxic
damage.

A neuroendocrine approach based on the measurement of serum prolactin (PRL) was identified as a possible way to confirm the extrapolation of neurotoxic effects affecting dopaminergic systems from experimental studies to human beings. In fact, PRL secretion is directly controlled by the activity of the tuberoinfundibular dopaminergic system (TIDA) exerting a tonic inhibition on pituitary lactotrope cells. Secreted PRL comes back to hypothalamic neurons in a short feedback loop and further stimulates TIDA activity. As a result, an increase in serum PRL gives rise to self inhibition through enhanced TIDA activity (Gudelsky, 1981). Increased serum PRL is a common finding among workers exposed to neurotoxic chemicals affecting dopaminergic systems. Such effects are detectable at exposure levels well below the current standards in industrialized countries. An example of risk characterization based on serum PRL will be discussed in the section on biomarkers and dose-response assessment (section II.6).

Validation of peripheral markers of neurotoxicity is still an exciting area of research, with little epidemiological and experimental evidence supporting the validity of peripheral markers of neurotoxicity.

II.4.3 Pneumotoxicity

So far, indeed, the assessment of health effects of pneumo-toxicants has mainly relied on such end-points as lung function impairment or respiratory symptoms which, although sensitive, do not permit an evaluation of oxidative damage caused by chemical exposure to the pulmonary epithelium. Recently, a new approach for assessing early effects of pollutants on the respiratory tract has been developed, based on the assay in serum of lung-derived proteins (Hermans & Bernard, 1998). One of these is the 15.8 kDa Clara cell protein (CC16), which is secreted in large amounts at the surface of airways from where it leaks into the serum most likely by passive diffusion. The serum concentration of CC16 is a new sensitive marker to detect an increased permeability of the epithelial barrier, which is one of the earliest signs of lung injury induced by air pollutants, including ozone. CC16 was measured in the serum of cyclists running for 2 h (between 2 and 4 p.m.) during summertime under different conditions, including episodes of photochemical smog, in Parma, Italy.

After the run, the serum concentration of CC16 was significantly increased and correlated with the ozone concentrations. The subjects were divided into quartiles of increasing ozone levels. Both pre- and post-run concentrations showed an exposure-related trend, the rise over the first quartile being significant from the fourth and third quartile onward, for the pre- and post-run, respectively (Fig. 10). Post-run CC16 in the fourth quartile was on average increased by 53%, as compared with the first quartile. When pre- and post-run values of serum CC16 were compared within each quartile by paired-samples Student's t test, the post-run elevation of serum CC16 was statistically significant already from the second quartile, corresponding to ozone levels between 0.060 and 0.084 ppm.

The ability of ozone to alter the lung epithelial barrier at ambient air levels was confirmed in animals. Female $C_{57}Bl/6$ mice were exposed to 0.08 ppm ozone for 4 and 8 h. As shown in Fig. 11, ozone produced an increase in serum CC16, statistically significant already after 4 h of exposure. After 8 h, this increase was more pronounced and accompanied by an influx of albumin and of polymorphonuclear neutrophils in bronchoalveolar lavage fluid (BALF). Thus, at this stage, the inflammatory response was associated with an enhanced bi-directional leakage of proteins across the pulmonary epithelial barrier which, by light microscopy, appeared morphologically intact.

The long-term significance of this altered epithelial permeability caused by ozone in ambient air is unknown. As shown by animal and human studies employing BALF, this phenomenon is a characteristic component of the acute inflammatory response to ozone that accompanies other inflammatory changes such as leukocyte influx and cytokine release. Of concern, several animal studies suggest that the prolonged maintenance of these effects might be detrimental to the lung tissue. In monkeys, exposure to 0.15 ppm ozone, 8 h a day for up to 90 days, resulted in morphological alterations of the pulmonary epithelium consisting of epithelium thickening and cellular proliferation in the interstitium. Similar epithelial lesions have been described in the lungs of rats exposed to an ozone concentration as low as 0.12 ppm, 12 h a day, for 6 weeks (Broekaert et al., 1999).

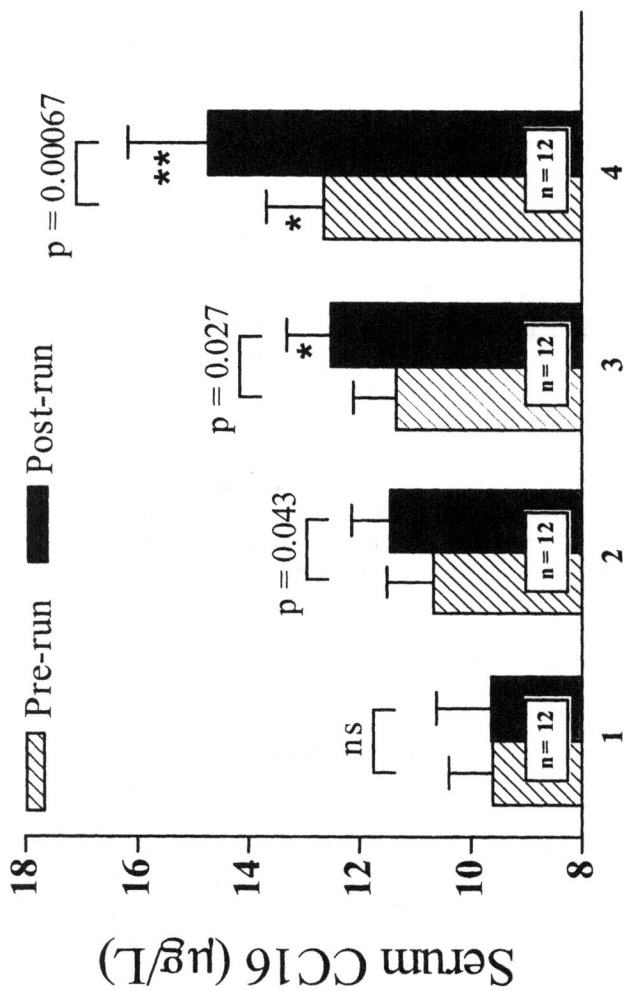

Fig. 10. Serum Clara cell protein (CC16, μg/litre) in cyclists before (pre-run) and after (post-run) a 2-h run in Parma (Italy). The pre-run serum concentrations of CC16 have been adjusted for that of cystatin C. Values of CC16 are given as mean ± SE. Quartiles correspond to the following ranges of O_3 concentrations: 0.0325–0.0595, 0.0605–0.084, 0.084–0.0925, 0.0925–0.103 ppm. Mean O_3 concentrations of the quartiles were 0.048, 0.072, 0.089 and 0.096 ppm, respectively. The p values refer to the comparison of pre- and post-run concentrations by the paired-samples Student's t test whereas asterisks indicate means which are significantly different from the first quartile (ns = not significant; * = P < 0.05; ** = P < 0.01; one-way ANOVA followed by Dunett's *post-hoc* test).

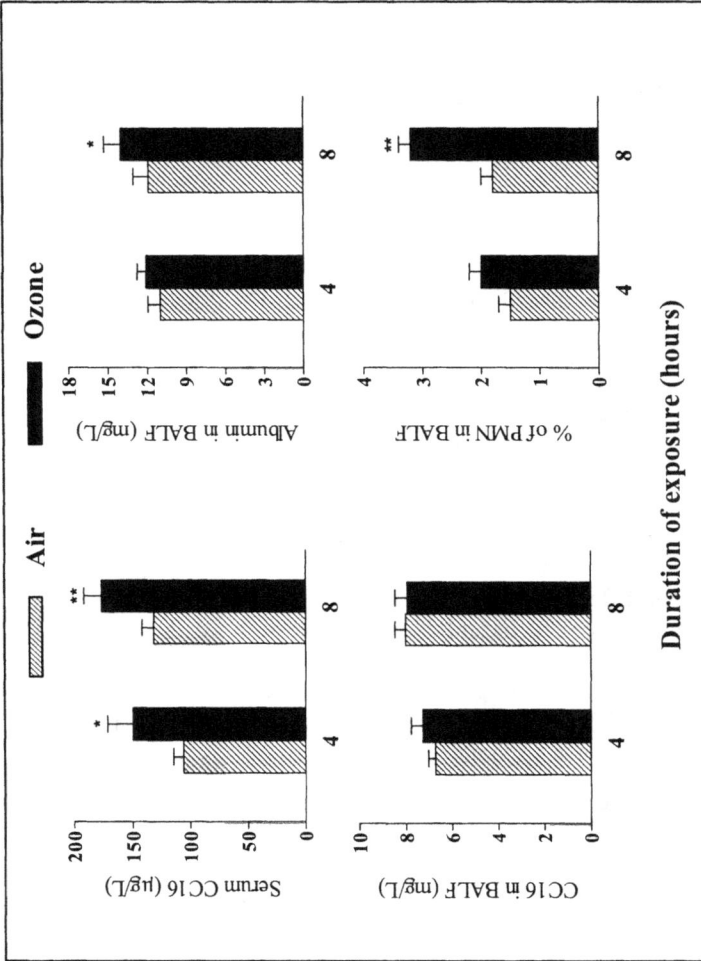

Fig. 11. Concentration of Clara cell protein (CC16, μg/litre) in serum and levels of CC16, albumin, and polymorphonuclear neutrophils (PMN) in bronchoalveolar lavage (BALF) from $C_{57}B1/6$ mice (Iffa Credo, L'Asbresle, France) exposed to 0.08 ppm O_3 or filtered air during 4 or 8 h. O_3 was produced from dried and filtered air by a high voltage generator (Anseros Ozomat, Anseros Klaus Nonnenmacher GMBH, Tübingen, Germany) and monitored every minute by a ultraviolet photometric analyzer (Signal Instrument Company, Farington-Oxon, UK). Bars represent mean ± SE of 6 to 7 animals (* = P < 0.05, Student's t test for independent samples).

125

II.4.4　Predictive validity of biomarkers of effect

Relevant markers applied in preventive programmes should meet several criteria, including non-invasiveness, sensitivity, predictive value for a possible evolution towards clinical disease, and some intrinsic health significance. Analytical methods must be reproducible, easy to perform and applicable to a large number of samples. Sampling procedures must be ethically acceptable, i.e., proportional to the importance of information for those undergoing the test.

II.4.4.1　*Methodological context*

Provided that the tests meet these general criteria, their use may require different features in terms of sensitivity and specificity, as well as practical feasibility, depending on the methodological context of their application. Basically, two types of preventive programmes can be envisaged: health surveillance and effect monitoring, the latter being usually carried out within the framework of epidemiological investigations. Both epidemiological studies (or effect monitoring) and health surveillance are applied to populations at risk, but the latter implies that an evaluation be made at the individual level. Since clinical diagnoses may have a considerable impact on the lifestyle and activities of individuals, specific rather than sensitive tests should be used and accuracy should be preferred to precision. Effect monitoring and epidemiological research are directed mainly toward hazard identification and risk assessment. These goals can only be achieved if the tests are sensitive enough to show early effects occurring at a time point when exposure to the suspected agent(s) is possible. In order to compare the test results over time, precision is preferable to accuracy.

Whereas reference values can be used to pick out individuals with abnormal values in screening procedures (health surveillance programmes), a reference group must always be concurrently used in effect monitoring and epidemiological investigations. If a group serves as its own control in a follow-up study, an internal quality control programme must ensure the comparability of results over time. Finally, whatever the approach, the markers must be relevant and meaningful. This means that they should actually measure what they are assumed to measure. This also means that unambiguous interpretation of the test results should be possible.

If a programme is effective, those who undergo it should have substantial benefits to their health in comparison to those who, under similar conditions, do not. It is conceivable that the earlier a marker is used, the higher is the likelihood that healthy subjects will be misclassified as ill people, which may have important consequences for the quality of their lives. Clinical investigations on individuals found to have markers should establish correct diagnoses and effective secondary prevention. Independently of the intrinsic properties of markers, the effectiveness of any preventive programme relies on its accuracy in targeting populations at risk, the predictive validity of any screening procedure being negatively related to the prevalence of the condition under consideration. In most screening programmes, a low prevalence of disease is the major cause of pitfalls. One means of increasing the effectiveness of individual evaluations is to use batteries of tests, which have opposite effects on group comparisons.

II.4.4.2 *Diagnostic validity*

The predictive value or diagnostic validity of a test is not a property of the test alone. It results from the combination of the sensitivity and specificity of the test and the prevalence of the disease in the population being examined. Positive results even for a very specific test, when applied to subjects with a low likelihood of being ill, will be largely false positives. In the clinical setting, where people are referred to because of symptoms/signs suggesting the occurrence of a disease, the positive predictive value of a diagnostic test (i.e., the probability that a subject with positive test results is actually ill) will be very high, whereas irrespective of the sensitivity and specificity, its negative diagnostic value will be low. Conversely, in field investigations the negative diagnostic value will be high and the positive diagnostic value will be low, no matter how sensitive and specific a test might be. As a result, most tests aimed at detecting early effects are useful to monitor health rather than to screen for illness. From the above paragraph, one might argue that biomarkers of effect are not useful because of their low positive diagnostic value. This is wrong for a number of reasons. It is worth mentioning that the low prevalence expected in field surveys refers to illness and not to early changes, which often occur at a relatively high rate. On the contrary, these changes are likely to occur, but in interpreting their significance the tendency to over-diagnose clinical diseases should be avoided.

II.4.4.3 Single biomarkers and test batteries

Whereas the very high prevalence of false-positive results obtained when screening for rare conditions may be reduced by confirmatory tests, the probability of finding statistically significant differences between groups when they do not actually occur (type 1 error) is clearly larger than the conventional 5% or 1% level in multigroup comparisons or when batteries of different tests have been used. It may be calculated that up to 40% of abnormal values may be anticipated when 10 independent tests are applied. In the case of renal markers, the tests are usually inter-related, which greatly reduces the probability of false positive results.

II.4.4.4 Prognostic value

A number of questions will arise when deviations from reference values are identified. The main question is a prediction regarding the outcome: prediction of the course of disease following its onset, prediction of the probability of illness following early changes, prediction that no adverse effects will be observed in future. Unfortunately, these predictions should be based on data that is usually not available. Although the questions are relevant, most often it is impossible to make meaningful predictions. When the risk and the prognosis are unknown, any indulgence in unwarranted conclusions should be avoided. Nevertheless, there are some public health issues that must be addressed. The first is the risk associated with early changes, i.e., the probability that early changes will be followed by a clinical disease. The second is the prognosis of observed changes, i.e., their consequences in terms of disability, complications or even death that may or may not follow our investigation. Owing to their low incidence in the general population, most degenerative diseases cannot be adequately studied using a prospective approach. In fact, their occurrence even in specific groups at risk will be extremely low. As a result, the use of markers sensitive enough to pick up early changes due to recent exposure seems now to be the only feasible approach to risk assessment. Obviously, the prognostic value of such early changes is often unknown. Nevertheless, it would be ethically unacceptable to "wait and see", when there are indications to improve a potentially harmful working environment (Mutti, 1993).

Experimental studies may be useful to validate biomarkers of effect when epidemiological investigations with sufficient power cannot be carried out. Some examples reported in this paper include the validation of indicators of early renal effect in rats exposed to hydrocarbons and the validation of early markers of lung toxicity in mice exposed to low ozone levels in ambient air.

II.5. PREDICTIVE VALIDITY AND RISK ASSESSMENT

In the use of biomarkers to establish dose-effect/response relationships, the most widely used epidemiological approach is that based on a cross-sectional study design. The rationale is that relevant markers of exposure and effect may be simultaneously measured and related to each other. Also, when examining a presumably healthy population to assess the risk of adverse effects, there is no reason to suspect a spontaneous selection of the sample, i.e., the most important bias in the analysis of prevalent data. Problems may arise when adverse health effects occur long after exposure took place or when the time-course of adverse health effects is unrelated to that of exposure (Becking, 1995). Limitations of the cross-sectional study as a risk-assessment method follow from its nature as an estimator of the point prevalence of adverse health effects, whereby the time sequence of exposure and effect cannot be established with certainty. Furthermore, a selection bias may represent a serious problem, since persons with certain illnesses or handicaps may be excluded from particular jobs or exposures, giving rise to a self-selection of the sample being surveyed, no longer representative of the whole population at risk. These biases make it difficult to separate causes from effects. Nevertheless, the cross-sectional approach based on the application of biomarkers is widely used in occupational toxicology to study dose-response relationships, because it allows the simultaneous measurement of both exposure and effects that are or may be attributable to exposure. If early markers are used, the above biases are unlikely to occur and the cross-sectional study design may provide the most relevant information for the prevention of possible long-term outcomes (Mutti, 1995). Some limitations are however apparent, especially in environmental health, since certain adverse effects may be detected only after maturation or differentiation to a stage at which the functioning of a system depends on the compromised structures. Toxic injury may also erupt when a second challenge unmasks some latent deficiency.

Furthermore, damage can be repaired and dysfunction compensated for, though sometimes it may trigger a process progressing to frank disease irrespectively of further exposure (Cone et al., 1987).

II.6. BIOMARKERS AND DOSE-RESPONSE ASSESSMENT

The definition of thresholds for safety purposes has been criticised, since no thresholds are evident for some toxicants (Bondy, 1985). Furthermore, any threshold is largely dependent on the sensitivity of the procedure used to measure adverse health effects, thus evolving with methodological progresses (Tilson, 1995). The threshold approach is also highly dependent upon the design of the study and the number of observations, i.e., upon the statistical power of the study. It is also worth mentioning that, when biomarkers are used to assess adverse effects, no clear cut-off points can be defined over the continuum linking early physiological changes to frank pathology. This is why the distinction between adverse and non-adverse effects is always debatable. Finally, even if reliable limit values were established, by definition they would not adequately consider the minority of hypersusceptible workers (Mutti, 1995).

In order to generate guidelines for occupational or environmental exposure on the basis of epidemiological data, dose-response relationships (in which the prevalence of abnormally high values is the relevant dependent variable) are more useful than dose-effect relationships (describing the increase in individual marker levels associated with increasing exposure). In the definition of thresholds, traditional approaches (NOAEL and LOAEL) require modifying and uncertainty factors, both are inappropriately related to sample size, and they may be inconsistent from study to study also because they are constrained to be an experimental or empirical dose. To counter objections to the NOAEL and LOAEL approach, the BMD approach has recently been proposed (Allen et al., 1994). The advantage of the BMD is that it better uses the information available rather than simply the lowest dose level at which effects are observed. The BMD is sensitive to the sample size, it can be applied consistently from one study to another, and it does not need to be an experimental or empirical dose level. This procedure has been applied to several non-cancer end-points.

In the benchmark dose approach, the prevalence of abnormalities is related to biomarkers of internal dose or to exposure levels, and the probability of adverse effects may be directly derived from the model describing such a relationship. In order to define a response, continuous variables must be transformed into nominal data or categories, i.e., into a simple dichotomy (for example, "abnormal" or "normal" levels). The location of a cut-off point distinguishing what is normal from what is abnormal in the distribution of a continuous variable is an arbitrary decision, usually based on statistical criteria, e.g., the 95th percentile of the reference interval (upper reference limit). The frequency of abnormal values (response) is related to dose levels using the mathematical model providing the "best fitting" of factual data. Since such a "response" occurs in the reference population, the threshold also has to be defined on the basis of statistical criteria. In environmental health, such a threshold is often defined as ED_{10}, i.e., the dose level corresponding to a two-fold increase over background occurrence (among the reference population or control subjects) or to a 5% excess risk. The benchmark dose (LED_{10}) is defined as the statistical lower bound on a dose corresponding to a specified level of risk - 10% or 5% excess risk in this case (Crump, 1984; Allen et al., 1994). From the upper confidence limit (95%) on the dose-response curve, the LED_{10} (lower confidence limit on the dose that produce a 10% risk) is obtained.

Fig. 12 illustrates the application of the benchmark dose approach to the risk of increased serum PRL on the basis of data recorded among 55 workers occupationally exposed to styrene. The cut-off had been defined as the 95th percentile of the PRL reference distribution (upper reference limit). The dose-response relationship and the upper 95% confidence limit of the logistic regression are plotted and the corresponding ED_{10}, and LED_{10} are identified (arrows). The dose-response curve fitted to available measurements gives rise to an ED_{10} of 40 mg of MAPGA per g creatinine ("next morning" spot sample). On the basis of exposure-dose relationships, such levels correspond to styrene concentrations of 5 ppm (8 h-TWA), i.e., to about 10% of the current TLV or about 25% of current exposure limits in some European countries based on neurobehavioral and genotoxic effects, respectively (IARC, 1993).

Styrene in air, 8h-TWA (ppm)

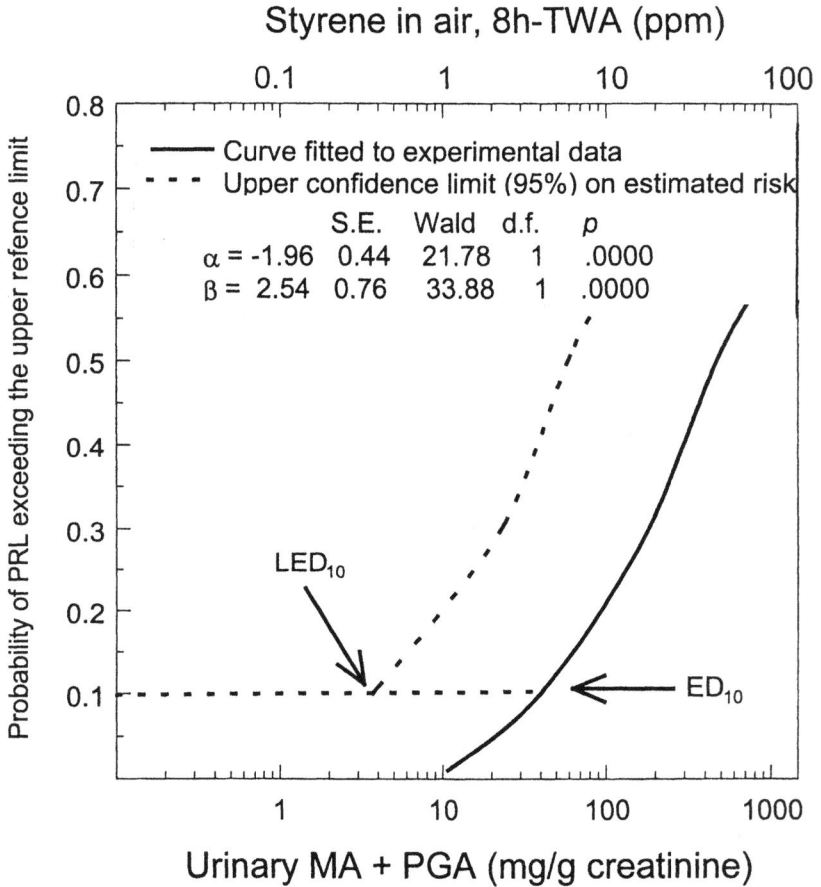

Fig. 12. Benchmark dose approach to risk assessment. The benchmark dose is defined as the statistical lower bound on a dose corresponding to a specified level of risk on the basis of the logistic function describing the dose-response curve between the urinary excretion of styrene metabolites in "next morning" spot samples (MAPGA) and the probability of increased serum prolactin. The ED_{10} (risk of 10% or excess risk of 5%) is estimated. From the upper confidence limit (95%) on the dose-response curve, the LED_{10} (lower confidence limit on the dose that produce a 10% risk) is obtained.

From the logistic regression model describing the dose-response curve, a LED_{10} (lower confidence limit on the dose associated with a 10% risk) of 4 mg/g creatinine is calculated. Such a level is below the detection limit for styrene metabolites as measured by HPLC or gas chromatography. The same procedure can be applied to available data collected among workers exposed to lead and manganese (Mutti & Smargiassi, 1998). The benchmark dose for styrene metabolites is about half the extrapolated NOAEL, whereas it is in the same order of magnitude for Mn-U, and about the double for Pb-B. Despite these variations, the benchmark dose and the extrapolated NOAEL for markers of internal dose appear to be in reasonable agreement. However, the benchmark dose is more informative, since it does not need the application of a default uncertainty factor, it incorporates the statistical uncertainty related to the data set used in the risk assessment, and it can provide a NOAEL on the basis of the logistic function describing the dose-response relationship even if actual data relevant to the point dose corresponding to the ED_{10} are not available.

A similar approach was used by Roels et al. (1993) to assess dose-response relationships between early markers of renal effect and urinary cadmium among occupationally exposed subjects (Fig. 13). Several dose-response curves could be identified depending on the marker and on the features of the examined population. As a general rule, dose-response assessment should be based on a broad dose-interval, which however tends to decrease over time as a result of improving working conditions. Such a situation will have even greater influence on dose-response assessment in future, when mobility across jobs and working environments will make it difficult to characterize exposure. Bernard & Lauwerys (1995) also drew attention on the fact that in the general population the tubulotoxic effects of cadmium may occur at a lower body burden of the metal than in adult male workers. Therefore, the application of an uncertainty factor would remain justified when extrapolating a no-effect level from adult male workers to the general population. It would be interesting to compare the BMD calculated from databases including occupationally and environmentally exposed subjects, to evaluate the influence of the exposure range as well as of other characteristics of each population on statistically defined thresholds.

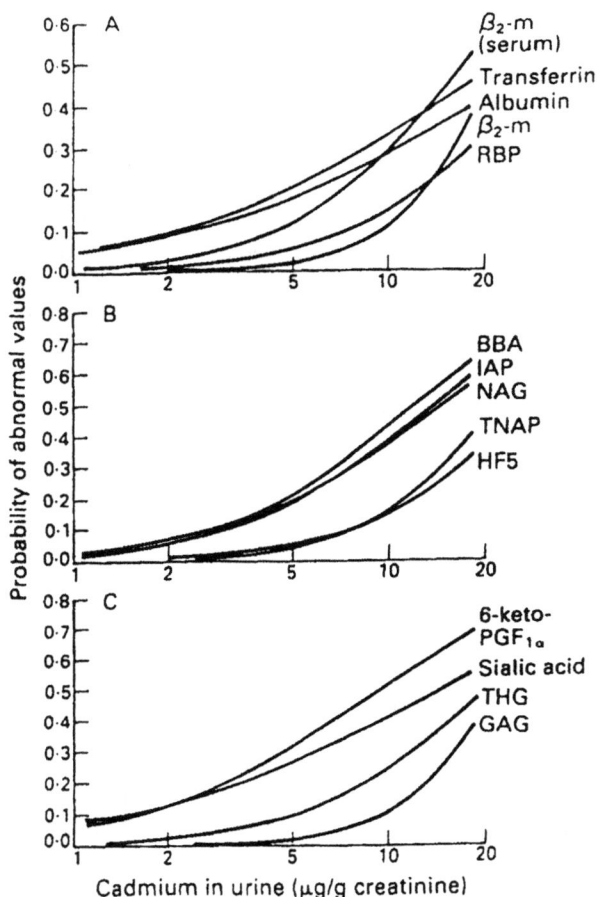

Fig. 13. Probability of renal effects as a function of urinary cadmium concentration (Cd-U) for (A) urinary proteins and serum β2-microglobulin, (B) enzymes or antigens in urine derived from the proximal tubule, and (C) other urinary markers. The upper limits of normal, defined as the 95th percentile of the values in control subjects with Cd-U lower than 1 μg Cd/g creatinine, were as follows) (see Table 11 for units): albumin, 19; transferrin, 903; β2-microglobulin, 324 (in serum: 2-17); RBP, 190; BBA, 6-7; IAP, 2-72; NAG, 2-19; TNAP, 1-9; HF5, 12-6; 6-keto-PGF$_{1x}$, 280; sialic acid, 501; THG, 50-1; and GAG, 63-2. The p values are: albumin O-114, transferrin 0-078, β2-microglobulin 0-099 (in serum 0-029), RBP 0-12, BBA 0-016, IAP 0-021, NAG 0-028, TNAP 0-068, HF5 0-094, 6-keto-PGF$_{1x}$, 0-016, sialic acid 0-052, THG 0-043, and GAG 0-099. All the markers of nephrotoxicity were standardized for the determinants unrelated to Cd exposure (see Table 10). Standardization was based on the mean of the total population.

The BMD approach can be applied to assess dose-response relationships between acrylamide adducts to haemoglobin (Hb) valine and neurological effects among workers from a Chinese chemical plant converting acrylonitrile to acrylamide. A logistic regression model based on published data (from Calleman et al., 1994) shows that 97.5% of subjects with neurological problems are correctly classified on the basis of acrylamide (AAVAL) and acrylonitrile (ANVAL) adducts to N-terminal valine in Hb. The corresponding benchmark dose in terms of acrylamide adducts to N-terminal valine in Hb is 0.8 nmol/g Hb (Fig. 14).

In summary, biomarkers of exposure and of effect can be concurrently used to assess dose-response relationships in populations at risk. Early and reversible effects among workers exposed to styrene vapours, such as increased serum prolactin, but also abnormalities depicted at neurobehavioral testing 15 h after the last exposure (Mutti et al., 1984b), show a dose-related trend when the urinary excretion of styrene metabolites is used as a marker of exposure (internal dose). Abnormalities of early markers of renal changes show dose-related trends when appropriate markers of body burden are used, e.g., urinary cadmium, whereas no dose-response relationship could be depicted when either air or blood concentrations are used to assess exposure to volatile organic solvents such as perchloroethylene (Mutti et al., 1992). Since consistent findings suggest a role of solvent exposure in the appearance of early renal changes and in their progression towards chronic renal disease (Mutti et al., 1996), one should conclude that such effects are either unrelated to dose levels or are related to exposure not accounted for by available markers of internal dose. Nevertheless, recent experimental evidence suggests that solvent exposure may have detrimental effects on kidney function and, especially, may accelerate the course of kidney disease from other causes (Mutti et al., 1999).

In the case of neurological abnormalities associated with exposure to acrylamide, adduct levels are related to the prevalence of abnormalities as assessed by neurophysiological and clinical criteria. Such a relationship is interesting, since, on the one hand, percutaneous absorption can occur among workers exposed to acrylamide. On the other, the biomarker of exposure is substance-specific, long-lived, and correlated with the clinical (and neuro-physiological) outcome.

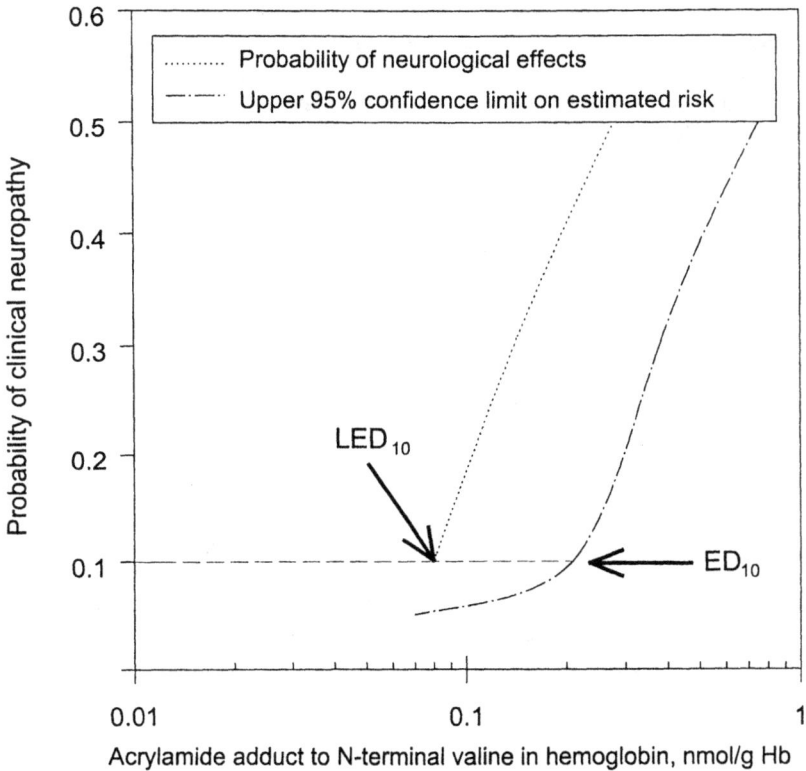

Fig. 14. Benchmark dose approach to risk assessment. The benchmark dose is defined as the statistical lower bound on a dose corresponding to a specified level of risk on the basis of the logistic function describing the dose-response curve between the concentration of acrylamide adducts to N-terminal valine in haemoglobin and the probability of abnormal neurotoxicity index, a composite score resulting from neuromyography, vibration threshold, and symptoms. The ED_{10} (risk of 10% or excess risk of 5%) is estimated. From the upper confidence limit (95%) on the dose-response curve, the LED_{10} (lower confidence limit on the dose that produce a 10% risk) is obtained. Calculated using data published by Calleman et al. (1994).

II.7. CONCLUSIONS

Biomarkers of effect can be used for screening purposes at the individual level, and to identify groups at risk in industry and in the general population. Whereas screening for disease is hampered by the prevalence dependence of the positive predictive value of diagnostic tests, the identification of groups at risk can rely on the shift in the distribution of sensitive markers of effect. Whether such a shift is sufficient to trigger action will always be debatable. In fact, the definition of biological limits or health-based criteria is hampered by: (i) the arbitrary nature of attempts to distinguish acceptable from unacceptable effects; (ii) the difficulties in assessing the prognostic value of observed changes; (iii) the influence of the study "power" on statistically defined thresholds. Whatever the approach, uncertainty will always characterize any scientific conclusion. This is why translating evidence into standards or even into broad categories (sufficient evidence, limited evidence, etc.) practical enough to implement regulations and effective risk management will always be questionable. However, the opposite is also true: there is little doubt about the possibility that the many thousand chemicals currently used may affect human health. This justifies a prudent attitude in interpreting early biological effects as a warning signal, requiring further support from mechanistic studies aimed at understanding their meaning and follow-up studies to confirm the existence of an increased risk for long-term outcomes (Mutti, 1995).

Unfortunately, consensus has not been reached as to whether risk assessment should rely on clinical criteria or on early markers of (adverse) effect. Furthermore, clinical criteria are sometimes used to question the validity of early markers as diagnostic tools, thus overinterpreting and misunderstanding their actual meaning as the forerunners of potentially serious health effect rather than as diagnostic tools. Such a biased perspective leads industrial organizations (e.g., ECETOC, 1996) to conclude that "there is no basis for a neurological syndrome in man that is causally related to organic solvents", and that "resources are better deployed on the control of exposure, rather than conducting further epidemiological investigations". Recent epidemiological research on Alzheimer's disease, which included also the most elderly, indicates that solvent exposure may be rather a strong risk factor (Kukull et al., 1995), and

Axelson (1996) concluded that "new clues suggesting ultimate causes or triggering mechanisms behind some of the serious neurological disorders may be obtained from occupational epidemiology". Obviously, there is no reason to believe that scientific research should lessen preventive efforts aimed at providing healthier workplaces. Ultimately, as recently stated (Mutti, 1995), "there are three basic questions that science and society must address before any action or even priority-setting may take place: (i) how much damage to our health are we prepared to tolerate?; (ii) how many (or what proportion of) affected individuals can we accept?; and finally, (iii) how much evidence is necessary before action is undertaken?

Once relevant end-points and criteria have been established, then suitable information can be properly used in formal risk assessment, bearing in mind that the most important cause of misclassification is misconception (Van Damme, 1996) and that the main cause of misinterpretation is misunderstanding. Biomarkers may be very useful to understand processes that are impossible to assess because of inherent difficulties.

Although the use of fully validated biomarkers is desirable, most often such a validation is largely incomplete. Validation in terms of testing the validity of assumptions generated by studies raising hypotheses is generally a difficult, painstaking process, based on converging evidence from multiple sources. In some cases, epidemiological investigations are not feasible and animal experiments can assist, but the intimate nature of biomarkers as surrogate measurements of something difficult or impossible to measure makes their full validation a quite unrealistic objective.

II.8. REFERENCES

Abbate M, Zoja C, Corna D, Capitanio M, Bertani T, & Remuzzi G (1998) In progressive nephropathies, overload of tubular cells with filtered proteins translates glomerular permeability dysfunction into cellular signals of interstitial inflammation. J Am Soc Nephrol, **9**: 1213-1224.

Aitio A & Kallio A (1999) Exposure and effect monitoring: a critical appraisal of their practical application. Toxicol Lett, **108**: 137-147.

Aitio A, Jarvisalo J, & Kiilunen M (1988) Chromium. In: Biological monitoring of toxic metals. Rochester Series on Environmental Toxicity. Eds.: Clarkson T, Friberg L, Nordberg G, Sager P, New York, Plenum Press.

Allen BC, Kavlock RJ, Kimmel CA, & Faustman EM (1994) Dose response assessments for developmental toxicity. II. Comparison of generic benchmark dose estimates with NOAELs. Fundam Appl Toxicol, **23**: 487-495.

Anger WK (1990) Worksite behavioural research: results, sensitive methods, test batteries and the transition from laboratory data to human health. Neurotoxicolgy, **11**: 629-720.

Anger WK (1993) Behavioural biomarkers to identify neurotoxic effects. In: Use of biomarkers in assessing health and environmental impacts of chemical pollutants, Edited by Travis CC, New York, Plenum Press.

Anger WK & Johnson BL (1985) Chemicals affecting behaviour. In: Neurotoxicology of industrial and commercial chemicals. O'Donoghue J, ed., CRC Press, Inc., Boca raton, FL.

Apostoli P (1997) Element speciation in biological monitoring [editorial]. Int Arch Occup Environ Health, **69**: 369-371.

Arfini G, Mutti A, Vescovi PP, Ferroni C, Giaroli C, Passeri M, & Franchini I (1987) Impaired dopaminergic modulation of pituitary secretion in workers occupationally exposed to styrene: Further evidence from PRL response to TRH stimulation. J Occup Med, **29**: 826-830.

Axelson O (1996) Where do we go in occupational neuroepidemiology? Scand J Work Environ Health, **22**(2): 81-83.

Becking GC (1995) Use of mechanistic information in risk assessment for toxic chemicals. Toxicol Lett, **77**: 15-24.

Bernard A (1995) Biokinetics and stability aspects of biomarkers: recommendations for application in population studies. Toxicology, **101**: 65-71.

Bernard AM & Lauwerys R (1995) Low-molecular-weight proteins as markers of organ toxicity with special reference to Clara cell protein. Toxicol Lett, **77**: 145-151.

Bernard A, Buchet JP, Roels H, Masson P, & Lauwerys R (1979) Renal excretion of proteins and enzymes in workers exposed to cadmium. Eur J Clin Invest, **9**: 11-22.

Bernard A, Broeckaert F, Hermans C, & Knoops B (1998) In Biomarkers: Medical and Workplace Applications, Mendelsohn ML, Mohr LC, & Peeters JP, Eds. Washington, DC, Joseph Henry Press, pp 273-283.

Bondy SC (1985) Especial considerations for neurotoxicological research. CRC Crit Rev Toxicol, **14**: 381-402.

Broeckaert F, Arsalane K, Hermans C, Bergamaschi E, Brustolin A, Mutti A, & Bernard A (1999) Lung epithelium damage at low levels of ambient ozone. Lancet, **353**: 900-901.

Calleman CJ, Wu Y, He F, Tian G, Bergmark E, Zhang S, Deng H, Wang Y, Crofton KM, Fennell T, & Costa LG (1994) Relationship between biomarkers of exposure and neurological effects in a group of workers exposed to acrylamide. Toxicol Appl Pharmacol, **126**: 361-371.

CEC/IPCS (1989) Health Significance and Early Detection of Nephrotoxicity. Toxicol Lett, **46**: 1-308.

Cone JE, Reeve GR, & Landrigan PJ (1987) Clinical and epidemiological studies. In: Toxic substances and human risk (Tardiff RG & Rodricks JV, eds.) New York, Plenum Press, pp 95-120.

Costa LG (1996) Biomarker research in neurotoxicology: the role of mechanistic studies to bridge the gap between the laboratory and epidemiological investigations. Environ Health Perspect, **104**(1): 55-67.

Costa LG & Manzo L (1995) Biochemical markers of neurotoxicity: research strategies and epidemiological applications. Toxicol Lett, **77**: 137-144.

Crump KS (1984) A new method for determining allowable daily intakes. Fundam Appl Toxicol, **4**: 854-871.

Droz P (1986) Simulation models for organic solvents. In: Safety and health aspects of organic solvents, Alan R. Liss, Inc., New York, pp 73-87.

Droz P (1993) Pharmacokinetic modelling as a tool for biological monitoring. Int Arch Occup Environ Health, **65**: 553-559.

ECETOC (1996) Chronic neurotoxicity of solvents, Technical Report No. 70, Brussels, European Centre for Ecotoxicology and Toxicology of Chemicals.

Ehrenberg L & Osterman-Golkar S (1980) Alkylation of macromolecules for detecting mutagenic agents. Teratog Carcinog Mutagen, **1**: 105-127.

Finley BL, Kerger BD, Katona MW, Gargas ML, Corbett GC, & Paustenbach DJ (1997) Human ingestion of chromium (VI) in drinking water, pharmacokinetics following repeated exposure. Toxicol Appl Pharmacol, **142**: 151-159.

Friberg L (1950) Health hazards in the manufacture of alkaline accumulators with special reference to chronic cadmium poisoning. Acta Med Scand, Suppl. 240.

Grandjean P (1995) Biomarkers in epidemiology. European Beckman Conference, Clin Chem, **41**(12): 1800-1803.

Grandjean P, Brown SS, Reavey P, & Young D (1995) Biomarkers in environmental toxicology. Clin Chem, **41**: 1799-1929.

Gudelsky, GA (1981) (1981) Tuberoinfundibular dopamine neurons and the regulation of prolactin secretion. Psychoneuroendocrinology. **6**: 3-16.

Hattis D (1996) Human interindividual variability in susceptibility toxic effects: from annoying detail to a central determinant of risk. Toxicology, **111**: 5-14.

Health and Safety Executive (1992) Biological monitoring for chemical exposures in the workplace. Guidance note EH 56 from the Health and Safety Executive. Her Majesty's Stationery Office, London.

Henderson RF, Bechtold WE, Bond JA, & Sun JD (1989) The use of biological markers in toxicology. Crit Rev Toxicol, **20**: 65-82.

Hermans C & Bernard A (1998) Lung epithelium specific proteins. Characteristics and potential applications as markers. Am J Resp Crit Care Med, **150** : 1-33.

Hulka BS (1990) Methodologic issues in molecular epidemiology. In: Hulka BS, Wilcosky TC & Griffith JD (eds.), Biological markers in epidemiology, New York, Oxford University Press, pp 214-226.

IARC (1993) Butadiene and Styrene: Assessment of Health Hazards. IARC Scientific Publications No. 127, International Agency for Research on Cancer, Lyon, 412 pp.

Ikeda M (1999) Solvents in urine as exposure markers. Toxicol Lett, **108**: 99-106.

IPCS (1991) Environmental Health Criteria 119: Principles and methods for the assessment of nephrotoxicity associated with exposure to chemicals. Geneva, World Health Organization, International Programme on Chemical Safety, 266 pp.

IPCS (1993) Environmental Health Criteria 155: Biomarkers and risk assessment: Concepts and principles. Geneva, World Health Organization, International Programme on Chemical Safety, 82 pp.

141

Kalliomaki P-L, Kalliomaki K, Rahkonen E, & Juntilla M-L (1985) Magnetopneumography-lung retention and clearance of manual metal arc welding fumes based on experimental and human data. In: Biomagnetism: Application and Theory. Eds.: Weinberg H, Stroink G, & Katila K, Oxford, Pergamon Press, p. 416.

Kukull WA, Larson EB, Bowen JD, McCormick WC, Teri L, & Pfanschmidt ML (1995) Solvent exposure as a risk factor for Alzheimer's disease: a case-control study. Am J Epidemiol, **141**: 1059-1071.

Lauwerys RR & Hoet P (1993) Industrial chemical exposure, guidelines for biological monitoring, 2nd edition, Boca Raton, Florida, Lewis Publishers.

Lison D, Buchet J-P, Swennen B, Molders J, & Lauwerys R (1994) Biological monitoring of workers exposed to cobalt metal, salts, oxide and hard metal dust. Occup Environ Med, **51**: 447-450.

Manzo L, Artigas F, Martinez E, Mutti A, Bergamaschi E, Nicotera P, Tonini M, Candura SM, Ray DE, & Costa LG (1996) Biochemical markers of neurotoxicity. A review of mechanistic studies and applications. Human Exp Toxicol, **15**(Suppl. 1): S20-S35.

Mercier MJ & Robinson AE (1993) Use of biologic markers for toxic end-points in assessment of risks from exposure to chemicals. Int Arch Occup Environ Health, **65**: S7-S10.

Mueller PW, Price RG, & Porter GA (1997) Proceedings of the Joint US/EU Workshop: Urinary biomarkers to detect significant effects of environmental and occupational exposure to nephrotoxins. Renal Failure, **19**(4): 501-504.

Mutti A (1991) Detection of renal disease in humans: developing markers and methods. Toxicol Lett, **46**: 177-192.

Mutti A (1993) Mechanisms and biomarkers of solvent-induced behavioral and neuroendocrine effects. In: Use of biomarkers in assessing health and environmental impacts of chemical pollutants (Travis CC ed.). New York, Plenum Publishing Corporation, pp 183-200.

Mutti A (1995) Use of intermediate end-points to prevent long-term outcomes. Toxicol Lett, **77**: 121-125.

Mutti A & Franchini I (1987) Toxicity of metabolites to dopaminergic systems and the behavioural effects of organic solvents. Br J Ind Med, **44**: 721-723.

Mutti A & Smargiassi A (1998) Selective vulnerability of dopaminergic systems to industrial chemicals: Risk assessment of related neuroendocrine changes. Toxicol Ind Health, **14**: 311-323.

142

Mutti A, Lucertini S, Fornari M, Franchini I, Bernard A, Roels H, & Lauwerys R (1984a) Urinary excretion of brush-border antigen revealed by a monoclonal antibody in subjects occupationally exposed to heavy metals. In: Proceedings of the 5th International Conference on Heavy Metals in the Environment. Edinburgh, CEP Ltd, pp 565-567.

Mutti A, Mazzucchi A, Rustichelli P, Frigeri G, Arfini G, & Franchini I (1984b) Exposure-effect and exposure-response relationships between occupational exposure to styrene and neuropsychological functions. Am J Ind Med, **5**: 275-286.

Mutti A, Pedroni C, & Arfini G (1984c) Biological monitoring of occupational exposure to different chromium compounds at various valence states. Int J Environ Anal Chem, **17**: 35-41.

Mutti A, Lucertini S, Valcavi P, Neri TM, Fornari M, Alinovi R, & Franchini I (1985) Urinary excretion of brush-border antigen revealed by a monoclonal antibody: early indicator of toxic nephropathy. Lancet, **ii**: 914-916.

Mutti A, Alinovi R, Bergamaschi E, Biagini C, Cavazzini S, Franchini I, Lauwerys RR, Bernard AM, Roels H, Gelpi E, Rosello J, Ramis I, Price RG, Taylor SA, De Broe M, Nuyts GD, Stolte H, Fels LM, & Herbort C (1992) Nephropathies and exposure to perchloroethylene in dry-cleaners. Lancet, **ii**: 189-193.

Mutti A, Bergamaschi E, Alinovi R, Lucchini R, Vettori MV, & Franchini I (1996) Serum prolactin in subjects occupationally exposed to manganese. Ann Clin Lab Sci, **26**: 10-17.

Mutti A, Coccini T, Alinovi R, Toubeau G, Broeckaert F, Bergamaschi E, Mozzoni P, Nonclercq D, Bernard A, & Manzo L (1999) Exposure to hydrocarbons and renal disease: an experimental animal model. Ren Fail, **21**(3-4): 369-85.

NRC (National Research Council) (1987) Biological markers in environmental health research. Environ Health Perspect, **74**: 3-9.

NRC (National Research Council) (1989a) Biological markers in reproductive toxicology, Washington, DC, National Academy Press, pp 395.

NRC (National Research Council) (1989b) Biological markers in pulmonary toxicology, Washington, DC, National Academy Press, pp 179.

NRC (National Research Council) (1992) Biological markers in immunotoxicology, Washington, DC, National Academy Press, pp 206.

NRC (National Research Council) (1995) Biological markers in urinary toxicology, Washington, DC, National Academy Press.

Nolan C (1995) Mechanistic basis for the development of biomarkers (summary report), Toxicol Lett, **77**: 81-83.

Piscator M (1962) Proteinuria in chronic cadmium poisoning. I. An electrophoretic and chemical study of urinary and serum proteins from workers with chronic cadmium poisoning. Arch Environ Health, 4: 607-621.

Piscator M (1966a) Proteinuria in chronic cadmium poisoning. III. Electrophoretic and immunoelectrophoretic studies on urinary proteins from cadmium workers, with special reference to the excretion of low molecular weight proteins. Arch Environ Health, 4: 335-344.

Piscator M (1966b) Proteinuria in chronic cadmium poisoning. IV. Gel-filtration and ion-exchange chromatography of urinary proteins from cadmium workers. Arch Environ Health, 4: 607-621.

Rahkonen E, Juntilla M-L, & Kalliomaki P (1983) Evaluation of monitoring among stainless steel welders. Int Arch Occup Environ Health, 52: 243.

Rappaport SM (1995) Biological monitoring and standard setting in the USA: a critical appraisal. Toxicol Lett, 77: 171-182.

Remuzzi G & Bertani T (1998) Pathophysiology of progressive renal disease. N Engl J Med, 339: 1448-1456.

Roels H, Bernard AM, Cardenas A, Buchet JP, Lauwerys RR, Hotter G, Ramis I, Mutti A, Franchini I, & Bundschuh I (1993) Markers of early renal changes induced by industrial pollutants. III. Application to workers exposed to cadmium. Br J Ind Med, 50: 37-48.

Savitz DA & Harlow SD (1991) Selection of reproductive health end-points for environmental risk assessment. Environ Health Perspect, 90: 159.

Schulte PA & Perera FP (1993) Validation. In: Schulte PA & Perera FP (eds.), Molecular epidemiology, San Diego, Academic Press, pp 79-107.

Shaikh ZA & Tohyama C (1984) Urinary metallothionein as an indicator of cadmium body burden and of cadmium induced nephrotoxicity. Environ Health Perspect, 54: 171-174.

Silbergeld EK (1993) Neurochemical approaches to developing biochemical markers of neurotoxicity: review of current status and evaluation of future prospects. Environ Res, 63: 274-286.

Tilson HA (1995) Defining neurotoxicity in a decision-making context. Neurotoxicol, 16: 363-376.

Travis CC (1993) Use of biomarkers in assessing health and environmental impacts of chemical pollutants, New York, Plenum Press, pp 284.

Van Damme (1996) Seminar: Current medican surveillance and pre-employment practices. Int J Occup Environ Health supplement to vol. S1-S61.

Welinder H, Littorin M, Gullberg B, & Skerkving S (1983) Elimination of chromium in urine after stainless steel welding. Scand J Work Environ Health, **9**: 397.

MEASUREMENT OF DRUG METABOLIZING ENZYME POLYMORPHISMS AS INDICATORS OF SUSCEPTIBILITY

Ari Hirvonen[1] & Olavi Pelkonen[2]

[1] Department of Industrial Hygiene and Toxicology, Finnish Institute of Occupational Health, Helsinki, Finland
[2] Department of Pharmacology and Toxicology, University of Oulu, Oulu, Finland

CONTENTS

III.1. INTRODUCTION

On the basis of epidemiological studies, up to 90% of all cancers are related to environmental factors, tobacco smoke and diet being the main attributable exposures (IARC, 1990). Lung cancer, for which tobacco smoke is unquestionably the most important causative factor, is currently the most common malignancy in the world. For other smoking-related cancers, about 50% of bladder cancers are attributable to smoking. Numerous chemical carcinogens are known to interact with cellular macromolecules and cause cancer initiation. Genetic host factors can interact with environmental carcinogens, i.e., carcinogens in the diet, tobacco smoke and ambient air due to environmental or occupational sources, and place an individual at a greater or lesser risk of a particular cancer than another individual (Raunio et al., 1995a,b; Hirvonen, 1999a,b).

The majority of carcinogens do not produce their biological effects *per se*, but require metabolic activation before they can interact with cellular macromolecules. Many compounds are converted to reactive electrophilic metabolites by the oxidative (phase I) enzymes, which are mainly cytochrome P450 enzymes (CYPs). Phase II conjugating enzymes, such as glutathione *S*-transferases (GST), UDP-glucuronosyltransferases (UGT) and *N*-acetyltransferases (NAT), usually act as inactivating enzymes.

A growing number of genes encoding phase I and II have been identified and cloned (Gonzalez, 1995; Nebert et al., 1996). Consequently, there is increasing knowledge of the allelic variants or genetic defects that lead to the observed variation. Development of fairly simple new techniques, such as PCR-based assays, has enabled identification of an individual's genotype for a variety of metabolic polymorphisms with precision. Thus, recent knowledge of the genetic basis for individual metabolic variation has opened new possibilities for studies focusing on increased susceptibility to environmental cancer. For extensive recent review articles, the reader could consult "Metabolic Polymorphisms and Susceptibility to Cancer" by Vineis et al. (1999) and articles by d'Errico et al. (1996), Wormhoudt et al. (1999) and Ingelman-Sundberg et al. (1999).

III.2. XME POLYMORPHISMS AND CANCER SUSCEPTIBILITY

III.2.1 Cytochromes P450

The cytochrome P450 (CYP)-dependent monooxygenases represent our first line of defence against toxic lipophilic chemicals by catalysing reactions involving the incorporation of an atom of molecular oxygen into the substrate (Guengerich, 1995). The resulting increase in hydrophilicity facilitates further metabolic processing and excretion. However, certain chemicals are activated to their ultimate carcinogenic form rather than being detoxified. Most carcinogen activation occurs through generation of epoxides or N-hydroxy intermediates that are further metabolized by transferases.

The main P450s in humans that metabolize carcinogens are CYP1A1, CYP1A2, CYP1B1, CYP2A6, CYP2E1, and CYP3A4 and CYP3A5. These enzymes have specificities for various classes of carcinogens, and genetic polymorphisms have been identified for most of them (Gonzalez et al., 1990, Guengerich, 1995). P450s are most extensively expressed in the liver although their levels of expression vary depending on the P450 form (Shimada et al., 1994). These inter-individual differences in expression may be due to genetic, host and environmental influences. Certain forms are also expressed in lung, gastrointestinal tract, kidney and larynx/nasopharyngeal tissue. In non-hepatic epithelial tissues, activation of carcinogens probably occurs directly in the cells being transformed, although in the case of arylamines and heterocyclic amines these chemicals are partially activated in the liver and transported to extrahepatic target sites where they undergo full activation (Kadlubar, 1994). Table 13 gives examples of CYP polymorphisms and their phenotypes.

I.2.1.1 CYP1A1 and CYP1A2

The CYP1A gene family has two members: CYP1A1, which is predominantly expressed in extrahepatic tissues such as the lung, and CYP1A2, which is concentrated in the liver (Nelson et al., 1996). CYP1A1 and CYP1A2 have overlapping catalytic activity and are both thought to play an important role in carcinogen activation.

Table 13. Major allelic variants and polymorphisms of human CYP genes and their phenotypic expressions[a]

Gene	Variant allele	In vitro	In vivo
CYP1A1	m1: 3'-Msp1-RFLP m2: Ile462Val 5' Xbal	no change no change mRNA expressed	no probe drugs known; lymphocytes, etc., as surrogates no change (see above) no change?
CYP1A2	not known	very variable	two-three phenotypes with probe drugs
CYP2A6	*2(v1): Lew160His *4: deletion	0 0	null phenotype null phenotype
CYP2B6	not known	variable (80% zero expression?)	no probe drugs known
CYP2C9	Ile359Leu Cys144Arg	decreased increased activity	retarded elimination of probe drugs (tolbutamide, warfarin) increased elimination of probe drugs (phenytoin, S-warfarin)
CYP2C19	m1: exon 5 cryptic splice site m2: exon 4 stop codon	0 0	retarded elimination or changed enantiomeric ratios of probe drugs (mephenytoin, omeprazole)

Table 13 (contd.)

CYP2D6	about 25 variant alleles identified	zero, increased, decreased	several useful probe drugs available
CYP2E1	5'-,3'-, intron-	no change?	probe drug chlorzoxazone
CYP2F1	not known	?	expression in lung
CYP3A4	not known	very variable	several probe drugs available (may measure also other CYP3A enzymes)
CYP3A5	not known	very variable	no specific probe drugs available
CYP3A7	not known	fetus-specific, very low in adults	no specific probe drugs available
CYP4B1	not known	?	expression in lung

[a] References are given in the text and the IARC Scientific Publication volume 148 (Vineis et al., 1999)

CYP1A1 is involved, for instance, in the metabolic activation of carcinogenic polycyclic aromatic hydrocarbons (PAHs), abundant in tobacco smoke, to their carcinogenic metabolites in the lung (Nelson et al., 1996). As an example, CYP1A1-dependent aryl hydrocarbon hydroxylase (AHH) activities in human lung tissue (microsomes) seem to be correlated to activation of benzo(a)pyrene-7,8-diol to the ultimate carcinogen (Rojas et al., 1992; Shou et al., 1996). Furthermore, the AHH activities are correlated with the benzo(a)pyrene-7,8-diol-9,10-epoxide (BaPDE) DNA-adduct levels in human lung tissue (Bartch et al., 1995).

The inter-individual variation in the CYP1A1-mediated AHH activity appears to have a yet unknown genetic basis. Using mitogen-stimulated peripheral blood mononuclear cells, Kellerman et al. (1973) observed a trimodal distribution of AHH induction, consistent with a codominant inheritance at a single genetic locus segregating for a more common allele conferring low inducibility, and a rarer allele conferring high inducibility. Later on, two closely linked genetic polymorphisms were detected within the CYP1A1 gene. The first polymorphism detected was a point mutation in the 3'-flanking region of the gene, creating a restriction fragment length polymorphism (RFLP) detected by MspI restriction enzyme (Kawajiri et al., 1990). Another polymorphic site has been found to be located in exon 7, where a nucleotide substitution causes an Ile to Val amino acid change in the haem-binding region of the enzyme (Hayashi et al., 1991a,b). Both the CYP1A1 Msp1 and Ile/Val variant alleles are much more prevalent in Asians than in Caucasians. More recently, two polymorphisms, one in exon 7 (Cascorbi et al., 1996a) and another in the 3' non-coding region (Crofts et al., 1994), have been identified. However, the effects of these genetic polymorphisms on CYP1A1 enzyme activity have thus far remained obscure (Landi et al., 1994; Crofts et al., 1994; Cascorbi et al., 1996a; Persson et al., 1997).

The expression of CYP1A1 is regulated by the cytoplasmic aryl hydrocarbon receptor (AHR), together with AHR nuclear translocator (ARNT) and several other regulatory proteins (Nebert, 1989; Swanson & Bradfield, 1993). Since no clear correlations have been observed between CYP1A1 allelic variants and lung cancer incidence in Caucasians, it has been suggested that variations in susceptibility to lung cancer may in fact be associated with polymorphisms in these genes affecting the CYP1A1 inducibility, rather than in the CYP1A1 gene itself.

Subsequent to the report suggesting that the extent of inducibility of CYP1A1 was increased in lymphocytes from lung cancer patients as compared to controls (Kellerman et al., 1973), a number of attempts have been made to confirm these findings (reviewed by d'Errico et al., 1996). Strong correlations between lung cancer risk and homozygosity for the CYP1A1 variant alleles have been reported in several Japanese studies (Kawajiri et al., 1990, 1993; Hayashi et al., 1991a,b; Sugimura et al., 1998). However, although similar association was also reported in an American population (Xu et al., 1996), no such association has been found in Europeans (Tefre et al., 1991; Hirvonen et al., 1992; Shields et al., 1992; Alexandrie et al., 1994; Bouchardy et al., 1997). The recent reports suggesting an association between CYP1A1 polymorphisms and increased risk of breast cancer (Ishibe et al., 1998) and endometrial cancer (Esteller et al., 1997) have not been substantiated (Dunning et al., 1999).

A number of studies have addressed the relationship between variant alleles and activity and/or inducibility, and have found either modest or no differences in CYP1A1-catalysed activities among the different alleles (Crofts et al., 1994; Landi et al., 1994; Wedlund et al., 1994; Jacquet et al., 1996; Persson et al., 1997). CYP1A1 is predominantly an extrahepatic enzyme and therefore it has been difficult to find an *in vivo* administered probe to determine its level of expression. In several studies surrogate tissues have therefore been used, which in most cases have been lymphocytes and monocytes, but hair follicles and some other easily available tissues have also been suggested for this purpose (see Raunio et al., 1995a,b). Critical research is needed in order to better define the genotype/phenotype relationships and also whether the activity and inducibility of CYP1A1 are critical for carcinogenicity.

CYP1A2 enzyme is one of the major P450s in human liver, representing on average about 15 % of the total P450 content (Guengerich, 1995). CYP1A2 metabolizes aflatoxin B1, various heterocyclic and aromatic amines, and certain nitroaromatic amines (Eaton et al., 1995). Several variant alleles at the 5'-region and intron 1 have been found in the Japanese population (Chida et al., 1999).

Their incidence in other ethnic groups and functional significance are not known, but considerable individual variations, both in the level of CYP1A2 expression in human liver (Ikeya et al., 1989) and in the rate of metabolism of CYP1A2 substrates, including aromatic amines, has been reported (Butler et al., 1992; Ilett et al., 1993; Eaton et al., 1995). CYP1A2 polymorphism may thus well be an important modifier of individual susceptibility to environmentally induced cancers.

II.2.1.2   CYP1B1

The CYP1B1 cDNA was cloned from keratinocyte cell lines on the basis of its inducibility by TCDD (Sutter et al., 1991). The gene is highly expressed in a multitude of human organs including kidney, brain, lymphocytes, endometrium, placenta, and fetal adrenal glands, lung, brain and kidney (Sutter et al., 1994; Hakkola et al., 1997), whereas its expression in liver is rather low. The CYP1B1 enzyme has been shown to catalyse the activation of several procarcinogens *in vitro* and therefore it was suggested that the enzyme may have a role in the carcinogenesis, particularly in extra-hepatic organs (Shimada et al., 1996a). From a cancer susceptibility point of view, another interesting substrate for CYP1B1 is estradiol. It seems probable that estradiol 4-hydroxylase is catalysed by CYP1B1 (Spink et al., 1994; Liehr et al., 1995).

The recent discovery of several mutations in the CYP1B1 gene and their possible role in congenital glaucoma (Stoilov et al., 1997) has also triggered interest in their possible functional consequences in terms of carcinogen activation and cancer susceptibility. There are some indications for the association between CYP1B1 poly-morphism and breast and lung cancer risk (Watanabe et al., 2000), but further studies about the functionatility of several variant alleles found (McLellan et al., 2000) and their significance for cancer risk are still needed for confirmation.

II.2.1.3   CYP2A6

CYP2A6 mediates the metabolism of several human carcinogens, including aflatoxins and nitrosamines. Phenotypic inter-individual and inter-ethnic variability of CYP2A6 activity seems to be wide, when measured by *in vivo* urinary excretion of

7-hydroxycoumarin or by coumarin 7-hydroxylation activity in liver samples (Cholerton et al., 1992; Rautio et al., 1992; Shimada et al., 1996b; Pelkonen et al., 2000). Originally, two variant alleles encoding inactive CYP2A6 (null alleles) were found (Fernandez-Salguero et al., 1995), but some problems in the original genotyping assay led to further studies and currently several variant alleles, including partial and total deletion alleles have been characterized (Oscarson et al., 1998, 1999a,b; Nunoya et al., 1998). Discrepant findings concerning possible associations between variant alleles and various cancers and diseases (pulmonary or hepatocellular carcinoma or liver cirrhosis) have been reported (Gullsten et al., 1997; London et al., 1999; Miyamoto et al., 1999). Since CYP2A6 is a principal enzyme metabolizing nicotine and cotinine (Messina et al. 1997), individuals carrying the CYP2A6-null alleles were recently suggested to have a decreased risk of developing tobacco-related cancers because they have a decreased risk of becoming addicted to smoking (Pianezza et al., 1998). Moreover, if they do become dependent, they seem to smoke less than those without impaired nicotine metabolism. As tobacco smoke contains nitrosamines, which can be activated to carcinogens by CYP2A6, these individuals may also be less efficient at activating the tobacco smoke-derived procarcinogens. However, these suggestions have not been confirmed.

I.2.1.4   CYP2C9 and CYP2C19

These enzymes are members of one of the most complex CYP families and both play a role in drug metabolism. Using *S*-mephenytoin as a probe drug, CYP2C19 was shown to exhibit polymorphic phenotype. The genetic basis for the deficiency in CYP2C19-mediated *S*-mephenytoin hydroxylation was elucidated first (Goldstein & de Morais, 1994) and the principal defect proved to be a single base pair mutation in exon 5 of the CYP2C19 gene, which creates an aberrant splice site (de Morais et al., 1994b). Another variant allele with a point mutation in exon 4 was also identified (de Morais et al., 1994a). The two variant alleles, originally named as m1 and m2, have been redesignated as CYP2C19*2 and CYP2C19*3, respectively, CYP2C19*1 being the wild-type allele (Daly et al.1996). CYP2C19*2 is much more common in Caucasians and CYP2C19*3 accounts for approximately 20% of the poor metabolizers in Orientals (de Morais et al.,

1994a,b). In addition to impaired *S*-mephenytoin metabolism, CYP2C19*2 homozygosity leads also to reduced omeprazole 5-hydroxylation (Ieiri et al., 1996).

The CYP2C9 gene also exhibits polymorphism. Three CYP2C9 alleles have thus far been identified (Daly et al., 1996), and expression studies of the variant alleles suggest that at least impaired warfarin and tolbutamide metabolism can be ascribed to these genetic variations (Haining et al., 1996; Sullivan-Klose et al., 1996).

Curiously, no procarcinogenic substrates are currently known for the CYP2C enzymes, but in view of the wide array of xenobiotics metabolized by members in this subfamily (Goldstein & de Morais, 1994; Gonzalez & Gelboin, 1994) the possibility of procarcinogenic substrates should not be ruled out. Two preliminary studies trying to link the CYP2C polymorphisms to cancer risk have been reported (London et al., 1996; Tsuneoka et al., 1996).

Along with CYP1A1 and CYP1A2, CYP2C9 also appears to play a role in the oxidative metabolism of BP. Allelic variants of CYP2C9 with functional repercussions have been identified (Rettie et al., 1994). Recently, a slight increased risk of lung cancer was associated with CYP2C9*2, which is the most common variant allele in Caucasians (London et al., 1997b), but contradictory findings have also been reported (Ozawa et al., 1997).

Inactive CYP2C19 alleles result in poor metabolism of *S*-mephenytoin, which has been shown to be more prevalent in Asians than Caucasians (Goldstein et al., 1997). The latter have approximately 1–2% poor metabolizers (PMs), while the former have up to 25% PMs. Interestingly, this is the opposite of the findings on CYP2D6 polymorphism.

II.2.1.5 CYP2D6

For CYP2D6, several reliable *in vivo* probes (debrisoquine, sparteine, dextromethorphan, metoprololol) are available and variant alleles or genomic changes leading to null, reduced or increased enzymatic activity have been extensively characterized (Caporaso et al., 1995; Daly et al., 1996). Individuals that are metabolically competent are referred as extensive metabolizers (EMs), and those that are incapable of metabolizing these drugs are PMs. Over 40 drugs are known to be substrates for CYP2D6 (Gonzalez, 1996). This polymorphism shows a marked ethnic difference in its frequency; 5–10% of Caucasians but < 1% of Asians lack expression of the active enzyme due to deficient CYP2D6 alleles. More than 10 variant alleles of the CYP2D6 that are partially or totally inactive have been characterized (Daly et al., 1996; Nelson et al., 1996).

The most common defective CYP2D6 allele among Caucasians is CYP2D6*4, which is characterized by a base substitution in the splice-site at intron 3/exon 4 boundary leading to a frame shift (Daly et al., 1996; Nelson et al., 1996). This allele was previously called CYP2D6B and accounts for more than 70% of all the inactivating alleles in Caucasian populations. Another variant allele, CYP2D6*3 (previously called CYP2D6A) consists of a single base pair deletion in the coding sequence in exon 5, also causing a frame shift. This allele accounts for about 5% of the alleles leading to loss of CYP2D6 enzyme activity (Daly et al., 1996; Nelson et al., 1996). The third loss of enzyme activity (~10–15% of the inactivating alleles) is caused by the deletion of the entire CYP2D6 gene (CYP2D6*5, previously called CYP2D6D). By analysing these three polymorphic sites, it is possible to identify at least 95% of European PMs (Broly et al., 1991; Daly et al., 1991). More recently a CYP2D6 allele representing amplification/duplication of the gene (CYP2D6*2XN) has been described (Johansson et al., 1993). Individuals who have inherited more than two copies of the CYP2D6 gene have been found to have a very high CYP2D6 enzyme activity and are consequently designated as ultrarapid metabolizers (UMs) (Meyer, 1994). The frequency of the duplicated allele seems to vary widely between populations of different ethnic origins. About 1% of Swedish, German, Chinese and black Zimbabwean populations are UMs (Masimirembwa et al., 1993; Johansson et al., 1994; Dahl et al., 1995; Sachse et al., 1997), whereas among Spaniards the frequency is 7% (Agundéz et al., 1995) and a very high prevalence has been observed among Saudi Arabians (21%; McLellan et al., 1997) and Ethiopians (29%; Aklillu et al., 1996).

Many studies have been performed on the potential association between polymorphic expression of CYP2D6 and the incidence of various types of cancer, with conflicting results (Smith et al., 1995; Nebert et al., 1996; d'Errico et al., 1996; Vineis et al., 1999).

However, the combined results of several studies carried out in various parts of the world suggest a significant but small decrease in the risk of lung cancer for individuals with CYP2D6 PM genotype (Rostami-Hodjegan et al., 1998). In keeping with this, an excess risk of lung cancer was recently associated with high CYP2D6 activity only in heavy smokers, a finding that may partly explain the inconsistent findings (Bouchardy et al., 1996).

II.2.1.6   CYP2E1

A few studies have been carried out which suggest that chlorzoxazone (6-hydroxylation) is a fair indicator of hepatic CYP2E1 activity (Peter et al., 1990; Dreisbach et al., 1995). At the genotypic level, several RFLP alleles (RsaI in the 5'-flanking region, DraI in intron 6 and TaqI in intron 7) have been uncovered (Ingelman-Sundberg & Johansson, 1995). Several studies have addressed the question whether chlorzoxazone metabolism *in vivo* is associated with the CYP2E1 variant alleles. So far, no associations have been found in either Caucasians or Japanese (Kim et al., 1995; Lucas et al., 1995; Kim et al., 1996; Carriere et al., 1996), indicating that the variant alleles found to date do not affect CYP2E1 activity *in vivo*.

In a Japanese study, individuals homozygous for the variant DraI alleles of CYP2E1 were reported to have decreased lung cancer risk especially among individuals with high cumulative smoking dose (Uematsu et al., 1992, 1994). In the Finnish population, this genotype was found to be much less frequent than in the Japanese population (Hirvonen et al., 1993). Moreover, no differences were observed in the frequency of this genotype between lung cancer patients and controls, in agreement with Swedish observations (Persson et al., 1997). Also the variant RsaI allele was found to be extremely rare among Finns and Scandinavians (Hirvonen et al., 1993; Persson et al., 1997). However homozygosity for the RsaI allele has been suggested to pose an increased risk of lung cancer in a Swedish study (Persson et al., 1997), while in Taiwanese this allele was associated with increased risk of nasopharyngeal carcinoma (Hildesheim et al., 1997).

II.2.1.7   CYP3A

Useful *in vivo* probes are available for the determination of CYP3A activity (midazolam, dapsone, erythromycin), (Wrighton &

Stevens, 1992), and several important carcinogens are known to be metabolized and activated by the CYP3A enzymes (Guengerich, 1994, 1995). Nevertheless, to our knowledge no studies have been published on the relationship between these enzymes as measured by *in vivo* probe drugs and cancer susceptibility. The probable reason for this is that only recently have CYP3A4 allelic variants been described. An allelic variant in the 5'-flanking region, although suggested to be associated with prostate cancer (Rebbeck et al., 1998), probably does not affect CYP3A4 functionally (Westlind et al., 1999). Two exon variants, which probably cause functional consequences, have not yet been studied with respect to cancer susceptibility (Sata et al., 2000).

III.2.2 Phase II enzymes

Polymorphisms of phase II genes, especially those concerning glutathione *S*-transferases and *N*-acetyltransferases, have been elucidated over the last 2 or 3 decades. Example of polymorphisms of phase II genes are shown in Table 14.

II.2.2.1 Epoxide hydrolase

Microsomal epoxide hydrolase (mEH) is an enzyme involved in the first-pass metabolism of highly reactive epoxide intermediates. It catalyses, with a broad substrate specificity, the conversion of highly reactive and cytotoxic arene oxides and aliphatic epoxides to less toxic trans-dihydrodiols (Oesch, 1973). The enzyme acts coordinately with, for example, CYP1A1 and CYP1A2 to inactivate deleterious polycyclic hydrocarbon oxides and epoxides. Further epoxidation can convert inactive diols to highly toxic, mutagenic and carcinogenic polycyclic hydrocarbon diol epoxides (Sims et al., 1974). Thus, epoxide hydrolase shows the same dual role of procarcinogen detoxification and activation found in some CYPs and, as a consequence, might also play an important role in epoxide toxicity.

The mEH enzyme is expressed in all tissues thus far examined (Oesch et al., 1977; Seidegård & Ekström, 1977), highest levels being in the liver, kidney, and testis and 10–100-fold lower levels in the lung and lymphocytes (Omiecinski et al., 1993). Within cells,

Table 14. Major polymorphisms of human phase II enzyme genes and their phenotypic expressions

Gene	Variant allele	*In vitro*	*In vivo*
EPHX	His₁₁₃Tyr	decreased	?
	Arg₁₃₉His	increased	?
GSTM1	gene deletion	0	probably null (Finns 50% n/n)
GSTM3	*B (deletion in intron 6)	?	?
GSTP1	*B (Ile104Val)	decreased	?
	*C (Ile104Val, Ala113Val)	decreased	?
GSTT1	gene deletion	0	probably null (Finns 15% n/n)
NAT1	*3 (C1095A)	?	?
	*10 (polyA signal change)	increased	?
	*11 (several changes)	?	?
	*14 (*10 + Arg187Gln)	decreased (absent?)	decreased
NAT2	> 23 mutations	decreased or absent	"traditional" slow acetylation phenotype

For references, see text and the IARC Scientific Publication volume 148 (Vineis et al., 1999).

mEH is localized mainly in the endoplasmic reticulum where it can transiently associate with the P450 system (Etter et al., 1991). Endogenous substrates for mEH have not been reliably identified. However, the high degree of mEH structural conservation between several mammalian species and apparent ubiquitous tissue expression implies an important role in cellular metabolism (Seidegård & Ekström, 1997).

Inter-individual differences in mEH activity ranging in scale from several to 40-fold have been reported in various human tissue types (Seidegård & Ekström, 1997). The molecular basis for variation in mEH activity has not yet been characterized completely. Genetic polymorphisms have, however, been identified within exons 3 and 4 of the mEH gene (EPHX; Hasset et al., 1994a,b), which result in $His_{113}Tyr$ and $Arg_{139}His$ amino acid substitutions, respectively. *In vitro* expression analyses indicated that the corresponding mEH activities are decreased by approximately 40% (Tyr_{113}) or increased by at least 25% (His_{139}). The activity level observed in the presence of both variations approximates that observed for the wild-type genotype (Hasset et al., 1994b). Recently a genetic variation in the 5' flanking sequence of EPHX was observed, which may be an additional contributing factor to the range of functional mEH expression existing in human populations (Raaka et al., 1998).

Data from the few studies addressing a possible association between EPHX polymorphisms and cancer support a dual role for the mEH in the carcinogenic process. The EPHX His_{113} variant allele has been suggested to increase the risk of aflatoxin-associated hepatocarcinoma (McGlynn et al., 1995) but to decrease the risk of ovarian cancer (Lancaster et al., 1996). With regards to lung cancer, no significant association was found to the EPHX genotypes (Smith & Harrison, 1997).

1.2.2.2 Glutathione S-transferases

Among the detoxification systems, the glutathione *S*-transferases (GSTs) play a critical role in providing protection against electrophiles and products of oxidative stress (Hayes et al., 1995). GSTs are a superfamily of enzymes having broad and overlapping substrate specificities; four families of cytosolic soluble GSTs have so far been identified in humans, referred to as Alpha, Mu, Pi and Theta (Hayes et al., 1995). The known substrates for GSTs in cigarette smoke are those derived from bioactivation of PAHs, namely, PAH diolepoxides. The most studied carcinogenic PAH diolepoxide, BPDE, is a good substrate for many GST isoforms like GSTM2, GSTM3 and especially for GSTP1 and GSTM1 (Coles & Ketterer, 1990; Hayes et al., 1995). In general, class Mu enzymes show highest activities with most epoxides.

To date, genetic polymorphism has been found in four of the GST genes. One of them is GSTM1, which is expressed in only about half of Caucasians, due to a homozygous deletion (null genotype) of the gene in the other half (Seidegård et al., 1988). In addition to the null genotype two functional alleles denoted as GSTM1*A and GSTM1*B have been described. These alleles differ by a base substitution ($C_{534}G$) in the latter, which has not been shown to affect the GSTM1 activity.

In several recent studies an increased risk of cancer has been observed among GSTM1 null smokers, but several conflicting reports also exist (London et al., 1995; McWilliams et al., 1995; d'Errico et al., 1996; Rebbeck, 1997). In the light of compiled data it has been estimated that 17% of both lung cancers (McWilliams et al., 1995) and bladder cancers (Brockmöller et al., 1994) may be attributable to GSTM1 genotypes. Although these values provide only a crude measure of the potential population impact of these genes, they suggest that GSTM1 deficiency could indeed contribute to a substantial fraction of cancer at the population level. In contrast, at the individual level the risk associated with the GSTM1 null genotype may be smaller than has been anticipated.

GSTM3 is one of the most abundant GSTs in human lungs (Inskip et al., 1995; Anttila et al., 1993, 1995). As a deviation from the wild-type GSTM3*A allele, the variant allele GSTM3*B carries a deletion of three base pairs in intron 6, which results in the generation of a recognition sequence for the YYI transcription factor. The functional consequence of this is still unclear, but both negative and positive regulatory effects have been suggested (Inskip et al., 1995; Yengi et al., 1996).

People with low expression of GSTM3 were previously observed to be at an increased risk of developing adenocarcinoma of the lung (Anttila et al., 1995). Recent genotyping studies have indicated that individuals who are homozygous or heterozygous for the GSTM3*B alleles would have a lower risk of cancers of the larynx (Jahnke et al., 1996) and lung (Matthias et al., 1998; Jourenkova et al., 1998) than those with homozygous wild-type genotype.

The third polymorphic GST gene, GSTP1, encodes an isoform that is known to metabolize many carcinogenic compounds, among them BPDE. Given that GSTP1 is the most abundant GST isoform in the lungs (Anttila et al., 1993), it is anticipated to be of particular importance in the detoxification of inhaled carcinogens. Two variant alleles, GSTP1*B and GSTP1*C, have been detected in addition to the wild-type allele GSTP1*A. GSTP1*B has an $A_{313}G$ transition in exon 5, causing an $Ile_{104}Val$ amino acid change. In addition to this base substitution, GSTP1*C allele has a $C_{341}T$ transition, resulting in a $Ala_{113}Val$ amino acid change. Both of the affected codons are in the electrophile-binding site of the GSTP1 enzyme (Ali-Osman et al., 1997). As compared to GSTP1*A, proteins encoded by GSTP1*B and GSTP1*C have been shown to have decreased enzyme activity when expressed in *Escherichia coli* (Zimniak et al., 1994; Ali-Osman et al., 1997). Individuals homozygous for the GSTP1*B alleles have been suggested to detoxify the ultimate carcinogen of BP, i.e., (+)-anti-BPDE, more efficiently than heterozygotes or wild-type homozygotes (Hu et al., 1997). Hence they could also be less susceptible to the carcinogenic effects of BP.

In a recent study, a three-fold increased risk of bladder and testicular cancer was observed for individuals homozygous for the GSTP1 low activity alleles (GSTP1*B and GSTP1*C alleles not differentiated) as compared to the controls (Harries et al., 1997). A similar association was also reported for cancers of the larynx (Matthias et al., 1998) and lung (Ryberg et al., 1997), followed by both supporting and contrasting findings (Ozawa et al., 1997; Jourenkova-Mironova et al., 1998; Harris et al., 1998).

A deletion polymorphism similar to that observed for GSTM1 has also been discovered for the GSTT1 gene (Pemble et al., 1994). The prevalence of GSTT1 null individuals shows a wide variation between ethnically different populations; in Caucasians the prevalence is 10–20% (Rebbeck, 1997). GSTT1 participates in detoxification of potentially carcinogenic monohalomethanes (Schröder et al., 1992) and of reactive epoxide metabolites of butadiene (Norppa et al., 1995; Wiencke et al., 1995), both of which are constituents of tobacco smoke. The GSTT1 null genotype has been associated with increased risk of lung (Saarikoski et al., 1998) and larynx cancers (Jourenkova et al., 1998), but similarly to the GSTM1 null genotype, controversial reports also exist (Deakin et al., 1996; To-Figueras et al., 1997; Kelsey et al., 1997).

Since different GST isoenzymes are known to have overlapping substrate specificities (Hayes & Pulford, 1995), deficiencies of GST isozymes may be compensated by other isoforms and utilization of alternative metabolic pathways. This may be one reason for the bulk of controversial data on GST polymorphisms and cancer proneness (Hirvonen, 1998, 1999a,b).

'II.2.2.3   N-*acetyltransferases*

Acetylation polymorphism (what is currently known as NAT2 polymorphism) was the first pharmacogenetic condition extensively studied (Weber & Hein, 1985). The *N*-acetylation polymorphism causes individual variations in biotransformation of various xeno-biotics with a primary aromatic amine or a hydrazine structure (Evans, 1992; Hein et al., 1993). The NAT2 (Blum et al., 1990), which was until recently thought to be the only polymorphic NAT, is responsible for the well-known inherited inter-individual variation in the ability to acetylate substrates such as the arylamine drugs procainamide and sulphamethazine, the arylamine carcinogen benzidine and some hydrazine drugs such as isoniazid and hydralazine (Evans, 1992; Hein et al., 1993). Recently another human *N*-acetyltransferase, NAT1 (Hein et al., 1993), which is widely expressed in tissues (Hearse & Weber, 1973) and in cultured cells (Coroneos & Sim, 1993), has also been found to be poly-morphic (Vatsis & Weber, 1993).

These findings may be of great clinical and toxicological importance since certain chemicals may be *N*-acetylated to a significant degree by both NAT1 and NAT2. These include the carcinogenic aromatic amines 2-aminofluorene; benzidine, 4-amino-phenyl, 4,4-dichloroaniline and 2-naphthylamine (Grant et al., 1991, 1992; Hein et al., 1992a,b; Lakshmi et al., 1995; Zenser et al., 1996), and the cancer chemotherapeutic agent dinaline (4-amino-*N*-[2'-aminophenyl] benzamide) (Grant et al., 1990). They are encoded at two distinct loci located on chromosome 8p21.3-23.1 along with NATP, a pseudogene, which does not encode a functional protein (Grant et al., 1997). The new nomenclature of NAT1 and NAT2 alleles is based on the consolidated classification system of Vatsis et al. (1995).

Seven NAT1 alleles in human populations have been reported in the literature so far (Grant et al., 1997). The NAT1*4 allele is denoted as the wild type. A prominent change in one of the variants

(NAT1*10), which possesses an alteration of the consensus polyadenylation signal (Vatsis & Weber, 1993), was recently reported to be associated with both higher NAT1 activity in bladder and colon tissue and DNA adduct levels in the colon tissues (Badawi et al., 1995; Bell et al., 1995). Given that NAT1 has been reported to be primarily responsible for the NAT activity in the human uroepithelium (Fredrickson et al., 1994), these findings are of special interest in studies on bladder cancer risk..The association between the NAT1*10 allele and NAT1 activity *in vivo* has not been confirmed in subsequent studies. This may at least partly be explained by previous misclassifications of a recently described NAT1*14 allele having $G_{560}A$ base substitution ($Arg_{187}Gln$), in combination with the $T_{1088}A$ and $C_{1095}A$ substitutions present in NAT1*10 allele. This allele produces a defective NAT1 protein, which leads to functional impairment in the metabolism of NAT1-selective substrates both *in vitro* and *in vivo* (Grant et al., 1997).

As for the NAT2 gene, in addition to the wild type allele NAT2*4, at least 23 different NAT2 mutations have been found to date (for references see Grant et al., 1997). Seven of the nine observed nucleotide transitions lead to amino acid changes while the remaining two base substitutions exert no influence on the amino acid sequence. Several allelic variants of NAT2 reported result from certain combinations of these nine base substitutions. Rapid acetylators have at least one wild-type NAT2*4 allele whereas slow acetylators have inherited two slow-acetylation-associated alleles.

Investigators have reported a wide range of values for acetylation activity in different groups (Lin et al., 1994). From the scarce population studies so far completed on NAT1, it appears that the NAT1 putative fast-acetylator alleles are found at a frequency ranging from 15–25% in Caucasians to 50% in Asians; NAT1*4 and NAT1*10 are the most prevalent alleles in Caucasians (Bell et al., 1995b; Probst-Hensch et al., 1996; Bouchardy et al., 1988). The predominance of the putative NAT1 slow-acetylator status-associated genotype (homozygous or heterozygous for NAT1*10) has been reported to be about 70% among British Caucasians (Bell et al., 1995b), 61% among French Caucasians (Bouchardy et al., 1998), and 50% among American population consisting of Caucasians, African Americans and Latinos (Probst-Hensch et al., 1996).

The NAT2 slow-acetylator alleles range from 5% in Japan to 90% in Egypt (Grant et al., 1997; Weber et al., 1988). The predominance of the NAT2 slow-acetylator genotype has been reported to be about 60% among Germans (Lin et al., 1993; Cascorbi et al., 1995), 53% among American Caucasians (Cascorbi et al., 1995), 63% among Polish (Mrozikiewicz et al., 1994) and 50% among Finns (Hirvonen et al., 1995). In contrast, in the Japanese or Chinese populations, the rapid genotype is largely over-represented (92% or 80%, respectively) (Mashimo et al., 1992; Rothman et al., 1993).

Previous phenotyping studies as well as subsequent genotyping studies have suggested a modifying role for NAT genotypes at all major cancer sites. Two main types of biological mechanisms could explain these findings (Hirvonen, 1999a,b). First, CYP-mediated N-hydroxylation of arylamines yields electrophilic intermediates, which are inactivated by conjugation with glucuronide or acetylation by NATs (Weber et al., 1988). In urinary bladder carcinogenesis, N-acetylation of arylamines represents a competing pathway for N-oxidation. The unconjugated N-hydroxy metabolites can enter the circulation, undergo renal filtration, and be transported to the urinary bladder (Kadlubar et al., 1992). A number of previous phenotyping studies provided evidence that the NAT2 slow-acetylator phenotype is a significant risk factor for the occurrence of bladder cancer, particularly in the case of occupational exposure to arylamines. Subsequent genotyping studies supported the important role of NAT2 slow-acetylation status as a risk factor for arylamine-induced bladder cancer (Weber et al., 1988; Risch et al., 1995; Golka et al., 1996). There is, however, also the possibility that slow acetylators survive longer than rapid acetylators with bladder cancer (Evans et al., 1983). Recent data suggest that a prominent variant allele of NAT1 (NAT1*10), associated with increased enzyme activity, is also a risk factor in smoking-related bladder cancer (Taylor et al., 1998).

Another line of research is based on the hypothesis that fast acetylators are at increased risk for cancers at other sites than bladder, due to the activation of procarcinogens such as heterocyclic amines. Exposure to heterocyclic amines is fairly common; these potent mutagens and rodent carcinogens are formed when meat and fish are cooked at household temperatures. The heterocyclic amines are poor substrates for *N*-acetylation in human liver, but they readily undergo hepatic *N*-oxidation and subsequent *N*-glucuronidation, resulting in conjugated *N*-hydroxy metabolites that can be transported to the colonic lumen (Turesky et al., 1991). In colonic mucosa, the *N*-hydroxy derivatives are good substrates for *O*-acetylation, which results in reactive *N*-acetoxyarylamines that are capable of forming covalent DNA adducts (Kadlubar et al., 1992). The association between the NAT1 fast-acetylator trait and colorectal tumours could be due to enhanced *O*-acetylation of aromatic amines in cigarette smoke or of heterocyclic amines in cooked meat, because both smoking and a high intake of red meat have previously been associated with colorectal cancer (Giovannucci & Willett, 1994; Giovannucci et al., 1994). The role of NAT1 activity is less clear if heterocyclic amines are the aromatic amine compounds of primary relevance to human colorectal cancer. Some data indicate that among the acetyltransferases, NAT2 is more important than NAT1 for bioactivation of heterocyclic amines *in vitro* (Minchin et al., 1992; Yanagawa et al., 1994; Wild et al., 1995; Yokoi et al., 1995).

Rapid acetylators were originally suggested to be at higher risk of developing cancer of the colon in several previous phenotyping studies. A couple of recent genotyping studies have reached a similar conclusion (Hirvonen, 1999a,b). Moreover, preliminary data suggest that the NAT1*10 allele is also a risk factor in smoking-related colon cancer (Bell et al., 1995b; Chen et al., 1998).

The *N*-acetylation phenotype has also been widely studied in relation to susceptibility to breast cancer and lung cancer. Several case-control studies compared the prevalence of the slow-acetylator phenotype in breast cancer patients with the prevalence found in controls, with mixed outcomes (Hirvonen, 1999a,b). Similarly, a recent genotyping study indicated an increased risk of breast cancer for slow NAT2 acetylators who smoked 20 or more cigarettes per day (Ambrosone et al., 1995), but two subsequent studies provided little evidence of an association between the NAT2 genotypes and breast cancer (Hunter et al., 1997; Millikan et al., 1998).

Other studies have evaluated the utility of acetylation as a risk marker for pulmonary malignancies and liver cancer. A set of four phenotyping studies yielded inconclusive results about the potential association between the NAT2 acetylator status and lung cancer risk (for review see Hirvonen, 1999a,b). In addition, the subsequent genotyping studies did not give any conclusive evidence (Martínez et al., 1995; Cascorbi et al., 1996b). However, the potential role of NAT genotypes as modifiers of individual responses to environmental agents was supported by three recent studies; the NAT2 slow acetylator genotype posed an increased risk of mesothelioma (Hirvonen et al., 1996) and hepatocellular carcinoma (Agúndez et al., 1996), while the NAT1 high activity allele increased the risk of smoking-related lung cancer (Bouchardy et al., 1998).

Evidently, *N*-acetylation may be an important detoxification step in environmental exposures. The combination of the NAT1 and NAT2 susceptible genotypes may appear to be a particularly unfavourable genotype composition in arylamine exposures. In agreement with this, the recently observed association between increased risk of colorectal cancer and the fast NAT1 acetylator allele (NAT1*10) was most apparent among fast NAT2 acetylators (Bell et al., 1995b). Moreover, this genotype combination, together with high red meat intake caused a remarkably increased colon cancer risk (Chen et al., 1998). Further addressing the potential importance of the individual acetylation capacity, the carcinogenic DNA adduct levels in the mucosa of the urinary bladder were found to be highest in arylamine-exposed individuals who had inherited both the slow NAT2 acetylator genotype and the rapid NAT1 acetylation-associated (NAT1*10) allele (Badawi et al., 1995).

II.2.2.4   *NAD(P)H:quinone oxidoreductase*

NAD(P)H:quinone oxidoreductase reduces quinones to dihydroquinones, a reaction that is considered to be critical in the detoxification of these highly reactive metabolites (Joseph et al., 1994). It is an important enzyme in both activation and detoxification pathways and is known to have a protective effect on the carcinogenicity and mutagenicity of quinone compounds and their metabolites and to activate procarcinogenic compounds (Lind et al., 1990). A polymorphic allele of the human NQO1 gene, with an amino acid change causing low catalytic activity (Traver et al., 1992; Eickelman et al., 1994; Rosvold et al., 1995), has recently been associated with increased susceptibility to malignancies such as colon and lung cancer (Marshall et al., 1991; Rosvold et al., 1995; Kolesar et al., 1995; Schulz et al., 1997).

III.2.3 Other potentially relevant XMEs

In addition to the abovementioned XMEs, a number of other polymorphic metabolic enzymes exist that may also be of importance in individual variations in susceptibility to cancer. Dihydropyrimidine dehydrogenase (DPD) is the first and rate-limiting enzyme in the catabolism of thymine and uracil. By virtue of its ability to degrade pyrimidines, DPD is responsible for the metabolism of fluoropyrimidine drugs, such as the extensively used cancer drug 5-fluorouracil (Gonzalez & Fernandez-Salguero 1995). DPD deficiency is associated with toxic effects after 5-fluorourasil treatment. An individual with a complete lack of DPD activity and his family were found to possess mRNA with a deletion, resulting in a non-functional DPD enzyme (Meisma et al., 1995). The reason for the deletion proved to be a G-to-A point mutation which leads to skipping of an entire exon during pre-RNA transcription (Wei et al., 1996). A genotyping test for this mutation is now available to detect individuals who would be at high risk of developing 5-fluorouracil toxicity during cancer chemotherapy (Wei et al., 1996).

Myeloperoxidase (MPO) is an enzyme found primarily in the lysosomes of neutrophils. Exposure to a variety of pulmonary insults, including cigarette smoke, stimulates recruitment of neutrophils into human lung tissue (Hunninghake et al., 1990) with local release of MPO (Schmekel et al., 1990a,b). MPO activates carcinogens in tobacco smoke including BP and aromatic amines (Mallet et al., 1991). An allelic variant with a G-to-A base substitution in the promoter region of the MPO gene has recently been shown to result in reduced gene transcription. Homozygotes for the variant allele have recently been suggested to be less susceptible to lung cancer (London et al., 1997a).

Sulfotransferases, which also comprise a superfamily, can participate in the metabolic activation of arylamine and heterocyclic amine carcinogens (Michedja et al., 1994). Although polymorphic variation of drug sulfation in humans has been quite extensively studied and characterized (Weinshilboum et al., 1997), there have been no attempts to our knowledge to link sulfation polymorphisms with cancer risk.

Thiopurine *S*-methyltransferase (TPMT) is a cytosolic enzyme that preferentially catalyses the *S*-methylation of aromatic and heterocyclic sulfhydryl compounds, including the commonly used cancer therapeutic agents mercaptopurine and thioguanine (Weinshilboum, 1992). PCR-based methods have been developed to detect the two variant alleles (TPMT*2 and TPMT*3), permitting diagnostics of heterozygous and homozygous individuals who would develop severe side effects if treated with standard doses of purine analogues (Krynetski et al., 1996).

The UDP-glycosyltransferases (UGTs), which also form a superfamily, conjugate active metabolites of carcinogens, and multiple forms are expressed in liver and extrahepatic tissues (Jin et al., 1993; MacKenzie et al., 1993; Babu et al., 1994). UGTs can also participate in the metabolism of arylamines and heterocyclic amines. While genetic polymorphisms of UGTs, especially those that conjugate bilirubin, have been described (Owens et al., 1995; Burchell et al., 1997), no studies on their possible association with cancer have been published.

The flavin-containing monooxygenases (FMOs) are a superfamily of xenobiotic-metabolizing enzymes that oxidize numerous nucleophilic compounds (Hines et al., 1994; Philips et al., 1995). These enzymes primarily carry out the inactivation of drugs and do not activate the common classes of carcinogens (Hines et al., 1994). A low-frequency polymorphism was found in FMO A1, that gives rise to a condition called trimethylaminurea or "Fish Odour Syndrome", which is due to the inability to carry out the *N*-oxidation of tertiary aliphatic amines found in foodstuffs (Philips et al., 1995).

The serum paraoxonase/acetylesterase catalyses the hydrolysis of organophosphate pesticides, such as paraoxon, carbamates and carboxylic acid esters. It also hydrolyses mustard gas and Sarin. A genetic polymorphism has also been found for this enzyme resulting in a high-activity and a low-activity allele (Furlong et al., 1993; Humbert et al., 1993).

III.2.4 Critical appraisal of the studies on XME polymorphisms and cancer susceptibility

The biological rationale in considering the activity and expression of xenobiotic metabolizing enzymes as possible cancer susceptibility factors is that by activating or inactivating carcinogenic substances the enzymes change the levels of carcinogens in the tissues and cells thus contributing to the multistage process of carcinogenesis at one or more stages. Hence, when considering the significance of phenotype and genotype in relation to cancer one obvious fact has to be stressed: it is the phenotype that is of importance to any possible outcome resulting from the exposure to chemicals. This fact being recognized, it becomes desirable that the first task is to investigate whether a genotypic change is actually "carried over" to the phenotype, i.e., what we can observe *in vivo* as a consequence of a chemical exposure. The necessity of elucidation of the relationship between genotype and phenotype is becoming particularly important nowadays when methods to uncover changes in the genomic DNA are rather easy, even routine. Modern molecular biological methods allow the determination of allelic variants at a fast rate, be they changes in single base pairs or in longer stretches of the DNA. In studies currently conducted, however, the problem is increasingly that the associations of allelic differences to cancer susceptibility are being explored without specific knowledge about whether the alleles under study are actually causing changes in the phenotype and whether the phenotype has even a theoretical association with the studied outcome.

Regulation of many of the XME genes is rather complex, with various environmental, host and genetic factors affecting the expression. It is therefore possible that in addition to mutations in the structural gene, mutations relevant to the phenotype may also occur in both 5' and 3'-flanking regulatory sequences and in other genes coding for transacting factors (e.g., regulatory proteins). Furthermore, gene deletion or multiplication may affect the enzymatic activity (for reviews see Meyer et al., 1990; Raunio et al., 1995; Kroemer & Eichelbaum, 1995; Fujii-Kuriyama et al., 1995; Ingelman-Sundberg & Johansson, 1995). For each of these categories several examples were presented in previous sections of this chapter: a number of single base-pair mutations in almost all the genes of carcinogen-metabolizing enzymes; a null allele of GSTM1 gene as an examples for structural gene deletion; a putative CYP1A1 regulatory gene (AHR gene) mutation affecting the inducibility of the CYP1A1 enzyme, and CYP2D6 gene multiplication.

It thus appears that several types of changes at the genomic level may influence the phenotype: 1) catalytic properties of an enzyme may be changed because protein structure is altered due to mutations in exons; 2) protein may altogether be absent due to mutations leading to truncated mRNA not capable of directing translation, or due to complete or partial deletion of the gene; or 3) protein may be expressed at variable levels because of changes in transcription or because of gene multiplication.

Different scenarios can be envisaged from simple cases to complex situations. The simplest case is "one gene-one protein", which solely catalyses the reaction of interest, and in which a single mutation leads to a complete inactivation of the gene product. The CYP2D6 or GSTM1 null alleles are examples of such cases. If the gene is not inducible or otherwise under remarkable environmental control, the correlation between genotype and phenotype should be, at least in principle, straightforward and simple. The situation is more complex when the reaction under study ("phenotype"), activation or inactivation of a carcinogen, is catalysed by a number of enzymes with variable catalytic properties and expressions and where the gene is under the influence of several environmental and host factors. In this situation the elucidation of the relationship between phenotype and genotype is extremely difficult. Unfortunately the situation often seems to be of the latter type, i.e., the phenotype-genotype relationship is very complex. A "complete" elucidation of the phenotype-genotype relationship is therefore possible only in a few cases where gene regulation is simple and probe substances that can be used are particularly informative for the catalytic properties of a single enzyme.

The extent of exposure may be very important for the outcome and for its dependence on genotypic/phenotypic traits (Vineis et al., 1994). It is quite possible that under exceptional, e.g., very high levels of, exposure conditions inter-individual differences in susceptibility are irrelevant, whereas they are of significance under low-level exposures typical for most human exposures. Thus the extent of exposure might be a modifying factor, which determines whether the role of a genotype/phenotype can be unravelled in a particular study.

III.2.5 Pros and cons of genotyping and phenotyping

I.2.5.1 *Genotyping*

It is clear that genotyping has some considerable advantages over phenotyping methods, such as unequivocal determination of the genetic background and lack of confounding by most host and environmental factors (Caporaso et al., 1995). There are, however, also some problems with genotyping studies. As already stated, the basic problem is to understand which phenotype is created by an allele. As recently as 5 years ago, this was in many cases a difficult task. With multiple alleles the task was even more daunting, at least when *in vivo* approaches were used, and often the results have been obtained from opportunistic studies (for example, availability of suitable individuals) rather than from well-defined experiments. At present there are some indirect methods to tackle this basic problem. With the aid of a variety of heterologous expression systems, it is possible, although sometimes technically demanding, to express variant alleles and measure the activities of the expressed proteins (Gonzalez & Korzekwa, 1995). If the activity in the *in vitro* system is known, it should be possible to design appropriate *in vivo* studies with suitable probe substances and thus obtain basic information about genotype-phenotype correlation. This information should be the basic building block in the validation of both genotyping and phenotyping assays.

I.2.5.2 *Phenotyping*

In the past, inter-individual metabolic variations have been determined by *in vivo* metabolism or combined *in vivo/in vitro* assays. Accordingly the phenotype has been determined in various ways, including direct measurements of enzyme activities in tissue samples (e.g., liver biopsies), *ex vivo* methods to measure inducibility in cultured lymphocytes or by using so-called probe drugs *in vivo* (Pelkonen & Breimer, 1994). Interpretation and reproducibility of these studies has often been difficult due to many confounding factors (Caporaso et al., 1995) and therefore all these phenotyping methods raise serious considerations.

2.5.2.1   In vivo *approaches: probe drugs*

A number of more or less suitable probe drugs have been employed for the determination of metabolic polymorphisms (Gonzalez, 1992; Pelkonen et al., 1995). Ideally the parameter to be determined, whether a metabolite, metabolic ratio or rate of elimination, should reflect quantitatively the expression of the allele that is being probed. In the simplest case, the metabolism of the probe drug is exclusively catalysed by only one enzyme encoded by one allele, and hence the polymorphism under study should basically be a differentiation between an active and an inactive (due to mutation or deletion) allele. In this case the use of a probe drug should give straightforward information about the genotype. Such a situation may exist, e.g., with some CYP2D6 alleles, which in the case of a homozygous individual may result in a lack of active enzyme and no debrisoquine 4-hydroxylase activity. If differences in the catalytic properties of enzymes encoded by the polymorphic alleles are only quantitative, the characteristics of the drug and the assay become crucial. In this case the ability to differentiate between two alleles giving rise to enzymes of different activities (or to differentiate between heterozygotes and homozygotes) becomes dependent on, for instance, the pharmacokinetic properties of the drug, the parameter to be measured and the exact time of blood or urine collection. For validation of the methodology in such situations, employment of theoretical calculations should be considered (see Tucker et al., 1995).

In addition to these principal problems, a number of practical issues should be addressed when probe drugs are used. Most of these issues are related to the administration of drugs or other chemicals to individuals: compliance with oral drugs taken without close supervision; possible risks due to potential side-effects and toxicity; inconvenience to both the researcher and the volunteer if a compound has to be given parenterally; consideration on the selection of route and dose, etc. (Pelkonen et al., 1995).

2.5.2.2   Ex vivo *approaches*

Sometimes the enzyme of interest is not significantly expressed in the tissue relevant for pharmacokinetics, which is normally liver (e.g., CYP1A1), or there are no appropriate *in vivo* probes available (e.g., for some GST enzymes). In these cases *in vivo* studies cannot be performed and the phenotype has to be characterized otherwise. One solution has been to use blood lymphocytes or monocytes as material for enzymatic determinations (Raunio et al., 1995b). These types of *ex vivo* assays may, however, add further complications in that the process of isolation and culturing cells may change the expression and activity of the enzyme to be measured. If the polymorphism of interest relates to "all or none" enzyme activity, this approach could be possible, but becomes difficult to use if quantitative differences ought to be analysed, as is the case when inducibility is being measured.

2.5.2.3   In vitro *approaches*

In humans, ethical considerations limit the availability of tissue samples, making it problematic to plan and execute systematic studies. In addition, there are several practical problems. Target tissues are available only occasionally and therefore conditions in which they can be obtained are opportunistic and may not be optimal for the preservation of enzyme activities, or the samples may not represent well the population relevant for the study. Because of these limitations, *in vitro* studies using tissue samples are useful mainly for basic investigations of phenotype-genotype correlations. In addition, liver biopsies may be very useful for the validation of *in vivo* probes since the pharmacokinetic parameters of a majority of probe drugs are determined by hepatic activities (Pelkonen & Breimer, 1994).

III.2.6   General problems of phenotyping

It is well documented that a number of host and environmental factors may influence and act as confounders of the drug metabolism phenotype. These include age, hormonal status, disease, drug-drug interactions and dietary habits (Kalow, 1993; Caporaso et al., 1995). While the studies have provided interesting information on the regulation of various enzyme activities and, in some cases, also on how the enzymatic activity could associate to cancer risk, it is difficult to judge from them which is the "real phenotype" that should be compared to the genotype. Since it is the gene that eventually dictates how the enzyme is expressed, even under the most complicated influence of endogenous and exogenous factors, it should theoretically be possible always to understand the relationship between the genotype and the phenotype. For this, a complete understanding of the gene regulation is, however, necessary and, except for a very few simple cases, we are at the moment very far from this goal.

Another problem, partly related to the confounding factors, is the timing of phenotype determination. For example, in practice it is very difficult to perform prospective studies where the phenotype is analysed at the same period as the exposure to (a) carcinogen(s). Usually phenotyping has to be performed in a case-control setting, i.e., in situations where the critical exposure could have taken place as long as tens of years earlier. In addition, an important problem is the stability of the phenotype over time. The phenotype at the time of measurement may not be the same as the phenotype at the onset of the disease, for instance, if the subject under investigation gives up smoking or changes diet.

One of the most challenging problems in the characterization of the human drug metabolism phenotype stems from the fact that most of the catalysing enzymes are expressed in a tissue-specific manner leading to great differences between tissues in the activation and inactivation of carcinogens. Therefore the phenotype of interest should, in principal, be determined in the tissue of interest. Because of the limited availability of human samples, this may, however, be possible only in rare cases, for example, occasionally in relation to surgery, and therefore it is difficult to design systematic studies on the assumption that tissue samples are available.

These problems have lead to various attempts to try to estimate the enzymatic activities in tissues of interest otherwise. One approach is to use indirect measurements, such as surrogate tissues or probe drugs *in vivo* to extrapolate the activity in the tissue of interest. Ideally, a surrogate tissue should represent the target tissue in such a way that the behaviour of a carcinogen, which is of importance for the final outcome, i.e., manifest cancer, is faithfully reproduced in (or is in a meaningful correlation with) the surrogate tissue. Unfortunately, in most instances, this prerequisite for the use of surrogate tissue cannot even be addressed, simply because we often have only limited information about the expression of relevant enzymes in the target tissue.

It is obvious therefore that ideal target tissues are difficult to obtain. Nevertheless, because of the difficulties in obtaining human material, surrogate tissues have been chosen for many studies in a more or less opportunistic way by using "whichever is available". Thus, samples such as blood lymphocytes, hair follicles, surface epithelia from skin or buccal mucosa, or "surplus" tissue from surgery have been used (for review see Raunio et al., 1995a,b). When using these tissues one should always keep in mind that they may not reflect well activities in the target tissue and, for instance, in *ex vivo* regulation studies their inducibility may be different from that of the target tissue.

In any case, human tissue samples are extremely valuable for basic studies when one wants to obtain information on the tissue specificity, catalytic properties and mode of regulation of human carcinogen metabolizing enzymes.

III.3. CONCLUDING REMARKS

It seems clear that we are still far from an adequate elucidation of the correspondence between phenotype and genotype in the case of most carcinogen-metabolizing enzymes. With the help of current molecular biological techniques, it is much easier to discover and characterize new variant alleles and genetic changes, and to employ them in straightforward genetic epidemiological studies, than to first proceed to more cumbersome studies to address the basic question as to how the genotype is determining the phenotype and whether there is any biologically plausible link to be expected between the genotypic differences and cancer susceptibility. Eventually, however, it would be necessary to achieve knowledge of the complete sequence of events, from the gene to the outcome, so that it could be clearly seen what are the implications and possible preventive and treatment strategies to be employed in those cases where clear associations between carcinogen-metabolizing enzymes and cancer susceptibility have been uncovered.

III.4 REFERENCES

Agundéz JAG, Ledesma MC, Ladero JM, & Benítez J (1995) Prevalence of CYP2D6 gene duplication and its repercussion on the oxidative phenotype in white population. Clin Pharmacol Ther, **57**: 265-269.

Agúndez JAG, Olivera M, Martínez C, Ladero JM, & Benítez J (1996) Identification and prevalence study of 17 allelic variants of the human NAT2 gene in a white population. Pharmacogenetics, **6**: 423-428.

Aklillu E, Persson I, Bertilsson L, Johansson I, Rodrigues F, & Ingelman-Sundberg M (1996) Frequent distribution of ultrarapid metabolizers of debrisoquine in an Ethiopian population carrying duplicated and multiduplicated functional CYP2D6 alleles. J Pharmacol Exp Ther, **278**: 441-446.

Alexandrie A-K, Ingelman-Sundberg M, Seidegård J, Tornling G, & Rannug A (1994) Genetic susceptibility to lung cancer with special emphasis on CYP1A1 and GSTM1: a study on host factors in relation to age at onset, gender and histological cancer types. Carcinogenesis, **15**: 1785-1790.

Ali-Osman F, Akande N, & Mao J (1997) Molecular cloning, characterization, and expression in *Escherichia coli* of full-length cDNAs of three human glutathione *S*-transferase Pi gene variants. Evidence for differential catalytic activity of the encoded proteins. J Biol Chem, **272**: 10004-10012.

Ambrosone CB, Freudenheim JL, Graham S, Marshall JR, Vena JR, Brasure JR, Michalek AM, Laughlin R, Nemoto T, Gillenwater KA, Harrington A, & Shields PG (1995) Cigarette smoking, *N*-acetyltransferase 2 genetic polymorphisms, and breast cancer risk. JAMA, **276**: 1494-1501.

Anttila S, Hirvonen A, Vainio H, Husgafvel-Pursiainen K, Hayes JD, & Ketterer B (1993) Immunohistochemical localization of glutathione *S*-transferases in human lung. Cancer Res, **53**: 5643-5648.

Anttila S, Luostarinen L, Hirvonen A, Elovaara E, Karjalainen A, Nurminen T, Hayes JD, Vainio H, & Ketterer B (1995) Pulmonary expression of glutathione *S*-transferase M3 in lung cancer patients: association with GSTM1 polymorphism, smoking, and asbestos exposure. Cancer Res, **55**: 3305-3309.

Babu SR, Lakshmi VM, Owens IS, Zenser TV, & Davis BB (1994) Human liver glucuronidation of benzidine. Carcinogenesis, **15**: 2003-2007.

Badawi A, Hirvonen A, Bell DA, Lang N, & Kadlubar FF (1995) Role of aromatic amine acetyltransferases NAT1 and NAT2, in carcinogen-DNA adduct formation in the human urinary bladder. Cancer Res, **55**: 5230-5237.

Barret JC (1995) Mechanisms for species differences in receptor-mediated carcinogenesis. Mutat Res, **333**: 189-202.

Bartsch H, Rojas M, Alexandrov K, Camus A-M, Castegnaro M, Malaveille C, Anttila S, Hirvonen A, Husgafvel-Pursiainen K, Hietanen E, & Vainio H (1995) Metabolic polymorphism affecting DNA binding and excretion of carcinogens in humans. Pharmacogenetics, **5**: S84-S90.

Bell DA, Badavi A, Lang N, Ilett KF, Kadlubar FF, & Hirvonen A (1995a) Polymorphism in the *NAT1* polyadenylation signal: association of *NAT1*10* allele with higher *N*-acetylation activity in bladder and colon tissue samples. Cancer Res, **55**: 5226-5229.

Bell DA, Stephens DA, Castranio T, Umbach DM, Watson M, Deakin M, Elder J, Hendrickse C, Duncan H, & Strange RC (1995b) Polyadenylation polymorphism in the acetyltransferase 1 gene (*NAT1*) increases risk of colorectal cancer. Cancer Res, **55**: 3537-3542.

Blum M, Grant DM, McBride W, Heim M, & Meyer UA (1990) Human arylamine *N*-acetyltransferase genes: isolation, chromosomal localization, and functional expression. DNA Cell Biol, **9**: 193-203.

Bouchardy C, Benhamou S, & Dayer P (1996) The effect of tobacco on lung cancer risk depends on CYP2D6 activity. Cancer Res, **56**: 251-253.

Bouchardy C, Wikman H, Benhamou S, Hirvonen A, Dayer P, & Husgafvel-Pursiainen K (1997) CYP1A1 genetic polymorphisms, tobacco smoking and lung cancer risk in a French Caucasian population. Biomarkers, **2**: 131-134.

Bouchardy C, Mitrunen K, Wikman H, Husgafvel-Pursiainen K, Dayer P, Benhamou S, & Hirvonen A (1998) *N*-acetyltransferase NAT1 and NAT2 genotypes and lung cancer risk. Pharmacogenetics, **8**(4): 291-298.

Brockmöller J, Kerb R, Drakoulis N, Staffeldt B, & Roots I (1994) Glutathione *S*-transferase M1 and its variants A and B as host factors of bladder cancer susceptibility: a case-control study. Cancer Res, **54**: 4103-4111.

Broly F, Gaedigk A, Heim M, Eichelbaum M, Morike K, & Meyer UA (1991) Debrisoquine/sparteine hydroxylation genotype and phenotype: Analysis of common mutations and alleles of CYP2D6 in a European population. DNA Cell Biol, **10**: 545-558.

Burchell B, McGurk K, Brierley CH, & Clarke DJ (1997) UDP-glucuronosyl-transferases. In: Sipes IG, Gandolfi AJ, & McQueen CA (eds) Comprehensive toxicology, Amsterdam, Pergamon Elsevier Science, **3**: 401-435.

Butler MA, Lang NP, Yong JF, Caporaso NE, Vineis P, Hayes RB, Teitel CH, Massengill JP, Lawsen MF, & Kadlubar FF (1992) Determination of CYP1A2 and NAT2 phenotypes in human population by analysis of caffeine urinary metabolites. Pharmacogenetics, **2**: 116-127.

Caporaso N, DeBaun MR, & Rothman N (1995) Lung cancer and CYP2D6 (the debrisoquine polymorphism): sources of heterogeneity in the proposed association. Pharmacogenetics, **5**: s129-s134.

Carriere V, Goasduff T, Ratanasavahn D, Morel F, Gautier JC, Guillouzo A, Beaune P, & Berthou F (1993) Both cytochromes P450IIE1 and 1A1 are involved in the metabolism of chlorzoxazone. Chem Res Toxicol, **6**: 852-857.

Carriere V, Berthou F, Baird S, Belloc C, Beaune P, & de Waziers I (1996) Human cytochrome P450 IIE1 (CYP2E1): from genotype to phenotype. Pharmacogenetics, **6**: 203-211.

Cascorbi I, Drakoulis N, Brockmöller J, Maurer A, Sperling K, & Roots I (1995) Arylamine *N*-acetyltransferase (NAT2) mutations and their allelic linkage in unrelated Caucasian individuals: correlation with phenotypic activity. Am J Hum Genet, **57**: 581-592.

Cascorbi I, Brockmöller J, & Roots I (1996a) A C4887A polymorphism in exon 7 of human CYP1A1: population frequency, mutation linkages, and impact on lung cancer susceptibility. Cancer Res, **56**: 4965-4969.

Cascorbi I, Brockmöller J, Mrozikiewicz PM, Bauer S, Loddenkemper R, & Roots I (1996b) Homozygous rapid arylamine *N*-acetyltransferase (NAT2) genotype as a susceptibility factor for lung cancer. Cancer Res, **56**: 3961-3966.

Chen J, Stampfer MJ, Hough HL, Garcia-Closas M, Willett WC, Hennekens CH, Kelsey KT, & Hunter DJ (1998) A prospective study of *N*-acetyl-transferase genotype, red meat intake, and risk of colorectal cancer. Cancer Res, **58**: 3307-3311.

Chida M, Yokoi T, Fukui T, Kinoshita M, Yokota J, & Kamataki T (1999) Detection of three genetic polymorphisms in the 5'-flanking region and intron 1 of human CYP1A2 in the Japanese population. Jpn J Cancer Res, **90**: 899-902.

Cholerton S, Idle ME, Vas A, Gonzalez FJ, & Idle JR (1992) Comparison of a novel thin layer chromatographic-fluorescence detection method with a spectrofluoromethric method for the determination of 7-hydrocoumarin in human urine. J Chromatogr, **575**: 325-330.

Coles B & Ketterer B (1990) The role of glutathione and glutathione transferases in chemical carcinogenesis. Crit Rev Biochem Mol Biol, **25**: 47-70.

Coroneos E & Sim E (1993) Arylamine *N*-acetyltransferase activity in human cultured cell lines. Biochem J, **294**: 481-486.

Costa M (1995) Model for the epigenetic mechanism of action of nongenotoxic carcinogens. Am J Clin Nutr, **61**: 666S-669S.

Crofts F, Taioli E, Trachman J, Cosma GN, Currie D, Toniolo P, & Garte SJ (1994) Functional significance of different human CYP1A1 genotypes. Carcinogenesis, **15**: 2961-2963.

Dahl M-L, Johansson I, Bertilsson L, Ingelman-Sundberg M, & Sjöqvist F (1995) Ultrarapid hydroxylation of debrisoquine in a Swedish population. Analysis of the molecular genetic basis. JPET, **274**: 516-520.

Daly AK, Armstrong M, Monkman SC, Idle ME, & Idle JR (1991) Genetic and metabolic criteria for the assignment of debrisoquine 4-hydroxylation (cytochrome P4502D6) phenotypes. Pharmacogenetics, **1**: 33-41.

Daly AK, Cholerton S, Armstrong M, & Idle JR (1994) Genotyping for polymorphisms in xenobiotic metabolism as predictor of disease susceptibility. Environ Health Perspect, **102**: S44-S61.

Daly AK, Brockmöller J, Broly F, Eichelbaum M, Evans WE, Gonzalez FJ, Huang J-D, Idle JR, Ingelman-Sundberg M, Ishizaki T, Jacqz-Aigrain E, Meyer UA, Nebert DW, Steen VM, Wolf CR, & Zanger UM (1996) Nomenclature for human CYP2D6 alleles. Pharmacogenetics, **6**:193-201.

Deakin M, Elder J, Hendrickse C, Peckham D, Baldwin D, Pantin C, Wild N, Leopard P, Bell DA, Jones P, Duncan H, Brannigan K, Alldersea J, Fryer AA, & Strange RA (1996) Glutathione S-transferase GSTT1 genotypes and susceptibility to cancer: studies of interactions with GSTM1 in lung, oral, gastric and colorectal cancers. Carcinogenesis, **17**: 881-884.

de Morais SMF, Wilkinson GR, Blaisdell J, Meyer UA, Nakamura K, & Goldstein JA (1994a) Identification of a new genetic defect responsible for the polymorphism of (S)-mephenytoin metabolism in Japanese. Mol Pharmacol, **46**: 594-598.

de Morais SMF, Wilkinson GR, Blaisdell J, Nakamura K, Meyer UA, & Goldstein JA (1994b) The major genetic defect responsible for the polymorphism of S-mephenytoin metabolism in humans. J Biol Chem, **269**: 15419-15422.

Dreisbach AW, Ferencz N, Hopkins NE, Fuentes MG, Rege AB, George WJ, & Lertora JJL (1995) Urinary excretion of 6-hydroxychlorzoxazone as an index of CYP2E1 activity. Clin Pharmacol Ther, **58**: 498-505.

Dunning AM, Healey CS, Pharoah PDP, Teare MD, Ponder BAJ, & Easton DF (1999) A systematic review of genetic polymorphisms and breast cancer risk. Cancer Epidemiol Biomarkers Prev, **8**: 843-854.

Eaton D, Gallagher EP, Bammler TK, & Kunze KL (1995) Role of cytochrome P4501A2 in chemical carcinogenesis: implications for human variability in expression and enzyme activity. Pharmacogenetics, **5**: 259-274.

d'Errico A, Taioli E, Chen X, & Vineis P (1996) Genetic polymorphisms and the risk of cancer: a review of the literature. Biomarkers, **1**: 149-173.

Eickelman P, Schulz WA, Rohde D, Schmitz-Drager B, & Sies H (1994) Loss of heterozygosity at the NAD(P)H:quinone oxidoreductase locus associated with increased resistance against mitomycin C in human bladder carcinoma cell line. Biol Chem Hoppe Seyler, **375**: 439-445.

Esteller M, Garcia A, Martinez-Palones JM, Xercavins J, & Reventos J (1997) Germ line polymorphisms in cytochrome-P450 1A1 (C4887 CYP1A1) and methylene tetrahydrofolate reductase (MTHFR) genes and endometrial cancer susceptibility. Carcinogenesis, **18**: 2307-2311.

Etter H, Richter C, Ohta Y, Winterhalter KH, Sasabe H, & Kawato S (1991) Rotation and interaction with epoxide hydrolase of P-450 in proteoliposomes. J Biol Chem, **266**:18600-18605.

Evans DAP, Eze LC, & Whitney EJ (1983) The association of the slow acetylator phenotype with bladder cancer. J Med Genet, **20**: 330-333.

Evans DA (1992) *N*-acetyltransferase. In: Kalow W (Ed.) Pharmacogenetics of drug metabolism. New York, Pergamon Press, pp 95-178.

Fernandez-Salguero P & Gonzalez FJ (1995) The CYP2A gene subfamily: species differences, regulation, catalytic activities and role in chemical carcinogenesis. Pharmacogenetics, **5**: S123-128.

Fernandez-Salguero P, Hoffman SMG, Cholerton S, Mohrenweiser H, Raunio H, Rautio A, Pelkonen O, Huang J, Evans WE, Idle JR, & Gonzalez FJ (1995) A genetic polymorphism in coumarin 7-hydroxylation: sequence of the human CYP2A genes and identification of variant CYP2A6 alleles. Am J Hum Genet, **57**: 651-660.

Fredrickson SM, Messing EM, Reznikoff CA, & Swaminathan S (1994) Relationship between *in vivo* acetylator phenotypes and cytosolic *N*-acetyl-transferase and *O*-acetyltransferase activities in human uroepithelial cells. Cancer Epidemiol Biomarkers Prev, **3**: 25-32.

Fujii-Kuriyama Y, Ema M, Mimura J, & Sogawa K (1994) Ah receptor: a novel ligand-activated transcription factor. Exp Clin Immunogenet, **11**: 65-74.

Fujii-Kuriyama Y, Ema M, Mimura J, Matsushita N, & Sogawa K (1995) Polymorphic forms of the Ah receptor and induction of the CYP1A1 gene. Pharmacogenetics, **5**: S149-S153.

Furlong CE, Costa LG, Hasset C, Richter RJ, Sundstrom JA, Adler DA, Disteche CM, Omiecinski CJ, Chapline C, & Crabb JW (1993) Human and rabbit paraoxonases: purification, cloning, sequencing, mapping and role of polymorphism in organophosphate detoxification. Chem Biol Interact, **87**: 35-48.

Giovannucci E & Willett WC (1994) Dietary factors and risk of colon cancer. Ann Med, **26**: 443-452.

Giovannucci E, Rimm EB, Stampfer MJ, Hunter D, Rosner B, Willett WC, & Speizer FE (1994) A prospective study of cigarette smoking and risk of colorectal adenoma and colorectal cancer in US men. J Natl Cancer Inst, **86**: 183-191.

Goldstein JA & de Morais SMF (1994) Biochemistry and molecular biology of the human CYP2C subfamily. Pharmacogenetics, **4**: 285-299.

Goldstein JA, Ishizaki T, Chiba K, De Morais SM, Bell D, Krahn PM, & Evans DA (1997) Frequencies of the defective CYP2C19 alleles responsible for the mephenytoin poor metabolizer phenotype in various Oriental, Caucasian, Saudi Arabian and American black populations. Pharmacogenetics, **7**: 59-64.

Golka K, Prior V, Blaszkewicz M, Cascorbi I, Schöps W, Kierfeld G, Roots I, & Bolt HM (1996) Occupational history and genetic N-acetyltransferase polymorphism in urothelial cancer patients of Leverkusen, Germany. Scand J Work Environ Health, **22**: 332-338.

Gonzalez FJ (1992) Human cytochromes P450: problems and prospects. Trends Pharmacol Sci, **13**: 346-352.

Gonzalez FJ (1995) Genetic polymorphism and cancer susceptibility: Fourteenth Sapporo Cancer Seminar. Cancer Res, **55**: 710-715.

Gonzalez FJ (1996) The CYP2D6 subfamily. In: Cytochromes P450s: metabolic and toxicologic aspects (Ioannides C, ed.). Boca Raton, FL: CRC Press, pp 183-210.

Gonzalez FJ & Gelboin HV (1994) Role of human cytochromes P450 in the metabolic activation of chemical carcinogens and toxins. Drug Metab Rev, **26**: 165-183.

Gonzalez FJ & Fernandez-Salguero P (1995) Diagnostic analysis, clinical importance and molecular basis of dihydropyrimidine dehydrogenase deficiency. Trends Pharmacol Sci, **16**: 325-327.

Gonzalez FJ & Korzekwa KR (1995) Cytochromes P450 expression systems. Annu Rev Pharmacol Toxicol, **35**: 369-390.

Gonzalez FJ, Aoyama T, & Gelboin HV (1990) Activation of promutagens by human cDNA expressed cytochrome P450s. Prog Clin Biol Res, **340B**: 77-86.

Grant DM, Vollmer K-O, & Meyer UA (1990) *In vitro* metabolism of dinaline and acetyldinaline by human liver. 12th European Workshop on Drug Metabolism. Basel, Switzerland, p 147.

Grant DM, Blum M, Beer M, & Meyer UA (1991) Monomorphic and polymorphic human arylamine N-acetyltransferases: a comparison of liver isozymes and expressed products of two cloned genes. Mol Pharmacol, **39**: 184-191.

Grant DM, Vohra P, Avis Y, & Ima A (1992) Detection of a new polymorphism of human arylamine N-acetyltransferase NAT1 using p-aminosalicylic acid as an *in vivo* probe. J Basic Clin Physiol Pharmacol, **3**: S244.

Grant DM, Hughes NC, Janezic SA, Goodfellow GH, Chen HJ, Gaedigk A, Yu VL, & Grewal R (1997) Human acetyltransferase polymorphisms. Mutat Res, **376**: 61-70.

Guengerich FP, Kim DH, & Iwasaki M (1991) Role of human cytochrome P450 IIE1 in the oxidation of many low molecular weight cancer suspects. Chem Res Toxicol, **4**: 168-179.

Guengerich FP (1994) Catalytic selectivity of human cytochrome P450 enzymes: relevance to drug metabolism and toxicity. Toxicol Lett, **70**: 133-138.

Guengerich FP (1995) Cytochromes P450 of human liver. Classification and activity profiles of the major enzymes. In: Pacifici GM & Fracchia GN (eds) Advances in drug metabolism in man. Luxembourg, Office for the Official Publications of the European Communities, p 179-231.

Gullsten H, Agundez JAG, Benitez J, Läärä E, Ladero JM, Diaz-Rubio M, Fernandez-Salguero P, Gonzalez F, Rautio A, Pelkonen O, & Raunio H (1997) CYP2A6 gene polymorphism and risk of liver cancer and cirrhosis. Pharmacogenetics, **7**: 247-250.

Haining RL, Hunter AP, Veronese ME, Trager WF, & Rettie AE (1996) Allelic variants of human cytochrome P450 2C9: baculovirus-mediated expression, purification, structural characterization, substrate stereoselectivity, and prochiral selectivity of the wild-type and I359L mutant forms. Arch Biochem Biophys, **333**: 447-458.

Hakkola J, Pasanen M, Pelkonen O, Hukkanen J, Evisalmi S, Anttila S, Rane A, Mäntylä M, Purkunen R, Saarikoski S, Tooming M, & Raunio H (1997) Expression of CYP1B1 in human adult and fetal tissues and differential inducibility of CYP1B1 and CYP1A1 by Ah-receptor ligands in human placenta and cultured cells. Carcinogenesis, **18**: 391-397.

Harries LW, Stubbins MJ, Forman D, Howard GCW, & Wolf R (1997) Identification of genetic polymorphisms at the glutathione S-transferase Pi locus and association with susceptibility to bladder, testicular and prostate cancer. Carcinogenesis, **18**: 641-644.

Harris MJ, Coggan M, Langton L, Wilson SR, & Board PG (1998) Polymorphism of the Pi class glutathione S-transferase in normal populations and cancer patients. Pharmacogenetics, **8**: 27-31.

Hasset C, Robinson KB, Beck NB, & Omiecinski CJ (1994a) The human microsomal epoxide hydrolase gene (EPHX1): complete nucleotide sequence and structural characterization. Genomics, **23**: 433-442.

Hasset C, Aicher L, Sidhu JS, & Omiecinski CJ (1994b) Human microsomal epoxide hydrolase: genetic polymorphism and functional expression in vitro of amino acid variants. Hum Mol Genet, **3**: 421-428.

Hayashi SI, Watanabe J, Nakachi K, & Kawajiri K (1991a) Genetic linkage of lung cancer-associated Msp I polymorphisms with amino acid replacement in the heme binding region of the human cytochrome P4501A1 gene. J Biochem, **110**: 407-411.

Hayashi S, Watanabe J, & Kawajiri K (1991b) Genetic polymorphisms in the 5'-flanking region change transcriptional regulation of the human cytochrome P450IIE1 gene. J Biochem (Tokyo), **110**: 559-565.

Hayes JD & Pulford DJ (1995) The glutathione S-transferase supergene family: Regulation of GST and the contribution of the isoenzymes to cancer chemoprotection and drug resistance. Crit Rev Biochem Mol Biol, **30**: 445-600.

Hearse DJ & Weber WW (1973) Multiple N-acetyltransferases and drug metabolism. Tissue distribution, characterization and significance of mammalian N-acetyltransferase. Biochem J, **132**: 519-526.

Hein DW, Rustan TD, & Grant DM (1992a) Human liver polymorphic (NAT2) and monomorphic (NAT1) N-acetyltransferase isozymes catalyze metabolic activation of N-hydroxyarylamine and N-hydroxy-N-acetyl-arylamine proximate carcinogens. FASEB J, **6**: A1274.

Hein DW, Rustan TD, Doll MA, Bucher KD, Ferguson RJ, Feng Y, Furman EJ, & Gray K (1992b) Acetyltransferases and susceptibility to chemicals. Toxicol Lett, **64/65**: 123-130.

Hein DW, Doll MA, Rustan TD, Gray K, Feng Y, Ferguson RJ, & Grant DM (1993) Metabolic activation and deactivation of arylamine carcinogens by recombinant human NAT1 and polymorphic NAT2 acetyltransferases. Carcinogenesis, **14**: 1633-1638.

Hennings H, Glick AB, Greenhalgh DA, Morgan DL, Strickland JE, Tennenbaum T, & Yuspa SH (1993) Critical aspects of initiation, promotion, and progression in multistage epidermal carcinogenesis. Proc Soc Exp Biol Med, **202**: 1-8.

Hildesheim A, Anderson LM, Chen C-J, Brinton LA, Daly AK, Reed CD, Chen I-H, Caporaso NE, Hsu M-M, Chen J-Y, Idle JR, Hoover RN, Yang C-S, & Chhabra SK (1997) CYP2E1 genetic polymorphisms and risk of nasopharyngeal carcinoma in Taiwan. J Natl Cancer Inst, **89**: 1207-1212.

Hines RN, Cashman JR, Philpot RM, Williams DE, & Ziegler DM (1994) The mammalian flavin-containing monooxygenases: molecular characterization and regulation of expression. Toxicol Appl Pharmacol, **125**: 1-6.

Hirvonen A (1997) Combinations of susceptible genotypes and individual responses to toxicants. Environ Health Perspect, **105**: 755-758.

Hirvonen A (1999a) Polymorphic NATs and cancer proneness. In: Metabolic polymorphisms and cancer. Boffetta P, Caporaso N, Cuzick J, Lang M, & Vineis P (eds.). IARC Scientific Publications No. 148, pp 251-270.

Hirvonen A (1999b) Polymorphisms of xenobiotic-metabolizing enzymes and susceptibility to cancer. Environ Health Perspect, **107**: 37-47.

Hirvonen A, Husgafvel-Pursianen K, Karjalainen A, Anttila S, & Vainio H (1992) Point mutational MspI and Ile/Val polymorphism linked in the CYP1A1 gene: lack of association with susceptibility to lung cancer in a Finnish population. Cancer Epidemiol Biomarkers Prev, **1**: 485–489.

Hirvonen A, Husgafvel-Pursiainen K, Anttila S, Karjalainen A, & Vainio H (1993) The human CYP2E1 gene and lung cancer: DraI and RsaI restriction fragment length polymorphisms in a Finnish study population. Carcinogenesis, **14**: 85-88.

Hirvonen A, Pelin K, Tammilehto L, Karjalainen A, Mattson K, & Linnainmaa K (1995) Inherited GSTM1 and NAT2 defects as concurrent risk modifiers for asbestos-associated human malignant mesothelioma. Cancer Res, **55**: 2981-2983.

Hirvonen A, Saarikoski S, Linnainmaa K, Koskinen K, Husgafvel-Pursiainen K, & Vainio H (1996) GST and NAT genotypes and asbestos-associated pulmonary disorders. J Natl Cancer Inst, **88**: 1853-1856.

Hu X, O'Donnel R, Srivastava SK, Xia H, Zimniak P, Nanduri B, Bleicher RJ, Awasthi S, Awasthi YC, Ji X, & Singh SV (1997) Active site architecture of polymorphic forms of human glutathione S-transferase P1-1 accounts for their enantioselectivity and disparate activity in the glutathione conjugation of 7,8-dihydroxy-9,10-oxy-1,8,9,10-tetrahydrobenzo(a)pyrene. Biochem Biophys Res Comm, **235**: 424-428.

Humbert R, Adler DA, Disteche CM, Hasset C, Omiecinski CJ, & Furlong CE (1993) The molecular basis of human serum paraoxonase activity polymorphism. Nat Genet, **3**: 73-76.

Hunninghake GW & Crystal RG (1990) Cigarette smoking and lung destruction: accumulation of neutrophils in the lungs of cigarette smokers. Ann Rev Respir Dis, **128**: 833-838.

Hunter DJ, Hankinson SE, Hough H, Gertig DM, Garcia-Closas M, Spiegelman D, Manson JE, Colditz GA, Willett WC, Speizer FE, & Kelsey K (1997) A prospective study of NAT2 acetylation genotype, cigarette smoking, and risk of breast cancer. Carcinogenesis, **18**: 2127-2132.

IARC (1990) Cancer: causes, occurrence and controls. Tomatis L, Aitio A, Day NE, Heseltine E, Kaldor J, Miller AB, Parkin DM, Riboli E, eds. International Agency for Research on Cancer, Lyon: IARC Scientific Publications No. 100 (1990).

Ieiri I, Kubota T, Urae A, Kimura M, Wada Y, Mamiya K, Yoshioka S, Irie S, Amamoto T, Nakamura K, Nakano S, & Higuchi S (1996) Pharmacokinetics of omeprazole (a substrate of 2C19) and comparison with two mutant alleles, CYP2C19m1 in exon 5 and CYP2C19m2 in exon 4, in Japanese subjects. Clin Pharmacol Ther, **59**: 647-653.

Ikeya K, Jaiswal AK, Owens RA, Jones JE, Nebert DW, & Kimura S (1989) Human CYP1A2 sequence, gene structure, comparison with the mouse and rat orthologous gene, and genetic differences in liver IA2 mRNA concentrations. Mol Endocrinol, 3: 1399-1408.

Ilett KF, Castleden WM, Vandongen YK, Stacey MC, Butler MA, & Kadlubar FF (1993) Acetylation phenotype and cytochrome P4501A2 phenotype are unlikely to be associated with peripheral arterial disease. Clin Pharmacol Ther, 54: 317-322.

Ingelman-Sundberg M (1993) Ethanol-inducible cytochrome P450 2E1. Regulation, radical formation and toxicological importance. In: Poli G, Albano E, & Dianzani MU (eds) Free radicals: From basic science to medicine. Birkhäuser, Basel, p 287-301.

Ingelman-Sundberg M & Johansson I (1995) The molecular genetics of the human drug metabolizing cytochrome P450s. In: Pacifici GM & Fracchia GN (eds) Advances in drug metabolism in man. Luxembourg, Office for Official Publications of the European Communities, p 534-585.

Ingelman-Sundberg M, Oscarson M, & McLellan RA (1999) Polymorphic cytochrome P450 enzymes: an opportunity for individualized drug treatment. Trends Pharmacol Sci, 20: 342-349.

Inskip A, Elexpuru-Camiruaga J, Buxton N, Dias PS, Macintosh J, Campbell D, Jones PW, Yengi L, Talbot JA, Strange RC, & Fryer AA (1995) Identification of polymorphism at the glutathione S-transferase, GSTM3 locus: evidence for linkage with GSTM1*A. Biochem J, 312: 713-716.

Ishibe N, Hankinson SE, Golditz GA, Spiegelman D, Willett WC, Speizer FE, Kelsey KT, & Hunter D (1998) Cigarette smoking, cytochrome P450 1A1 polymorphisms, and breast cancer risk in the Nurses' Health Study. Cancer Res, 58: 667-671.

Jackson MA, Stack HF, & Waters MD (1993) The genetic toxicology of putative nongenotoxic carcinogens. Mutat Res, 296: 241-277.

Jacquet M, Lambert V, Baudoux E, Muller M, Kremers P, & Gielen J (1996) Correlation between P450 CYP1A1 inducibility, MspI genotype and lung cancer incidence. Eur J Cancer, 32A: 1701-1706.

Jahnke V, Matthias C, Fryer A, & Strange R (1996) Glutathione S-transferase and Cytochrome-P-450 polymorphism as risk factors for squamous cell carcinoma of the larynx. Am J Surg, 172: 671-673.

Jaskula-Sztul R, Reinikainen M, Husgafvel-Pursiainen K, Szmeja Z, Szyfter W, Szyfter K, & Hirvonen A (1998) Glutathione S-transferase M1 and T1 genotypes as and susceptibility to smoking-related larynx cancer. Biomarkers, 3: 149-155.

187

Jin CJ, Miners JO, Burchell B, & MacKenzie PI (1993) The glucuronidation of hydroxylated metabolites of benzo(a)pyrene and 2-acetylaminofluorene by cDNA expressed human UDP-glucuronocyltransferases. Carcinogenesis, **14**: 2637-2639.

Johnson BS, Brooks BA, Reyes H, Hoffman EC, & Hankinson O (1992) An MspI RFLP in the human ARNT gene, encoding a subunit of the nuclear form of the Ah (dioxin) receptor. Hum Mol Genet, **1**: 351.

Johansson I, Lundqvist E, Bertilsson L, Dahl M-I, Sjöqvist F, & Ingelman-Sundberg M (1993) Inherited amplification of an active gene in the cytochrome P450 CYP2D locus as a cause of ultrarapid metabolism of debrisoquine. Proc Natl Acad Sci, **90**: 11825-11829.

Johansson I, Oscarson M, Yue Q-Y, Bertilsson L, Sjöqvist F, & Ingelman-Sundberg M (1994) Genetic analysis of the Chinese cytochrome P450D locus: characterization of variant CYP2D6 genes present in subjects with diminished capacity for debrisoquine hydroxylation. Mol Pharmacol, **46**: 452-459.

Joseph P, Xie T, Xu Y, & Jaiswal AK (1994) NAD(P)H:quinone oxidoreductase 1 (DT-diaphorase): expression, regulation and role in cancer. Oncol Res, **6**: 525-532.

Jourenkova N, Reinikainen M, Bouchardy C, Dayer P, Benhamou S, & Hirvonen A (1998) Larynx cancer risk in relation to glutathione S-transferase M1 and T1 genotypes and tobacco smoking. Cancer Epidemiol Biomarkers Prev, **7**: 19-23.

Jourenkova-Mirnova N, Wikman H, Bouchardy C, Voho A, Dayer P, Benhamou S, & Hirvonen A (1998) Role of glutathione S-transferase GSTM1, GSTM3, GSTP1, and GSTT1 genotypes in modulating susceptibility to smoking related lung cancer. Pharmacogenetics, 8(6): 495-502.

Kadlubar FF, Butler MA, Kaderlik KR, Chou HC, & Lang NP (1992) Polymorphisms for aromatic amine metabolism in humans: relevance for human carcinogenesis. Environ Health Perspect, **98**: 69-74.

Kadlubar FF (1994) Biochemical individuality and its implications for drug and carcinogen metabolism: recent insights from acetyltransferase and cytochrome P4501A2 phenotyping and genotyping in humans. Drug Metab Rev, **26**: 37-46.

Kalow W (1993) Pharmacogenetics: its biologic roots and the medical challenge. Clin Pharmacol Ther, **54**: 235-241.

Kalow W & Tang B-K (1993) The use of caffeine for enzyme assays: a critical appraisal. Clin Pharmacol Ther, **53**: 503-514.

Kawajiri K, Nakachi K, Imai K, Yoshii A, Shinoda N, & Watanabe J (1990) Identification of genetically high risk individuals to lung cancer by DNA polymorphisms of the cytochrome P450IA1 gene. FEBS Lett, **263**: 131-133.

Kawajiri K, Nakachi K, Imai K, Watanabe J, & Hayashi S (1993) The CYP1A1 gene and cancer susceptibility. Crit Rev Oncol Hematol, **14**: 77-87.

Kawajiri K, Watanabe J, Hidetaka E, Nakachi K, Kiyohara C, & Hayashi S (1995) Polymorphisms of human Ah receptor gene are not involved in lung cancer. Pharmacogenetics, **5**: 151-158.

Kellermann G, Shaw CR, & Luyten-Kellerman M (1973) Aryl hydrocarbon hydroxylase inducibility and bronchogenic carcinoma. N Engl J Med, **289**: 934-937.

Kelsey K, Spitz MR, Zuo ZF, & Wiencke JK (1997) Polymorphisms in the glutathione S-transferase class mu and theta genes interact and increase susceptibility to lung cancer in minority populations. Cancer Causes Control, **8**: 554-559.

Ketterer B, Harris JM, Talaska G, Meyer DJ, Pemble SE, Taylor JB, Lang NP, & Kadlubar FF (1992) The glutathione S-transferase supergene family, its polymorphism, and its effects on susceptibility to lung cancer. Environ Health Perspect, **98**: 87-94.

Kim RB & O'Shea D (1995) Interindividual variability of 6-hydroxylation of chlorzoxazone in men and women and its relationship to CYP2E1 genetic polymorphisms. Clin Pharmacol Ther, **57**: 645-655.

Kim RB, O'Shea D, & Wilkinson GR (1994) Relationship in healthy subjects between CYP2E1 genetic polymorphisms and the 6-hydroxylation of chlorzoxazone: a putative measure of CYP2E1 activity. Pharmacogenetics, **4**: 162-165.

Kim RB, Yamazaki H, Chiba K, Oshea D, Mimura M, Guengerich FP, Ishizaki T, Shimada T, & Wilkinson GR (1996) *In vivo* and *in vitro* characterization of CYP2E1 activity in Japanese and Caucasians. J Pharmacol Exp Ther, **279**: 4-11.

Kinzler KW & Vogelstein B (1996) Life (and death) in a malignant tumour. Nature, **379**: 19-20.

Kolesar JM, Kuhn JG, & Burris HA III (1995) Detection of a point mutation in NQO1 (DT-diaphorase) in a patient with colon cancer. J Natl Cancer Inst, **87**: 1022-1024.

Kroemer HK & Eichelbaum M (1995) Molecular bases and clinical consequences of genetic cytochrome P450 2D6 polymorphism. Life Sci, **56**: 2285-2298.

Krynetski EY, Tai H-L, Yates CR, Fessing MY, Loennechen T, Schuetz JD, Relling MV, & Evans WE (1996) Genetic polymorphism of thiopurine S-methyltransferase: clinical importance and molecular mechanisms. Pharmacogenetics, **6**: 279-290.

Lakshmi VM, Bell DA, Watson M, Zenser TV, & Davis BB (1995) *N*-acetylbenzidine and *N,N'*-diacetylbenzidine formation by rat and human liver slices exposed to benzidine. Carcinogenesis, **16**: 1565-1571.

Lancaster JM, Brownlee HA, Bell DA, Futreal A, Marks JR, Berchuck A, Wiseman RW, & Taylor JA (1996) Microsomal epoxide hydrolase polymorphism as a risk factor for ovarian cancer. Mol Carcinogenesis, **17**: 160-162.

Landi MT, Bertazzi PA, Shields PG, Clark G, Lucier GW, Garte SJ, Cosma G, & Caporaso NE (1994) Association between CYP1A1 genotype, mRNA expression and enzymatic activity in humans. Pharmacogenetics, **4**: 242-246.

Liehr JG, Ricci MJ, Jefcoate CR, Hannigan EV, Hokanson JA, & Zhu BT (1995) 4-hydroxylation of estradiol by human uterine myometrium and myoma microsomes: implications for the mechanism of uterine tumorigenesis, Proc Natl Acad Sci USA, **92**: 9220-9224.

Lin H, Han C-Y, Lin BK, & Hardy S (1993) Slow acetylator mutations in the human polymorphic *N*-acetyltransferase gene in 786 Asians, Blacks, Hispanics, and Whites: application to metabolic epidemiology. Am J Hum Genet, **52**: 827-834.

Lin H, Han C-Y, Lin BK, & Hardy S (1994) Ethnic distribution of slow acetylator mutations in the polymorphic *N*-acetyltransferase (NAT2) gene. Pharmacogenetics, **4**: 125-134.

Lind C, Cadenas E, Hochstein P, & Ernster L (1990) DT-diaphorase: purification, properties and function. Methods Enzymol, **186**: 287-301.

London SJ, Daly AK, Cooper J, Navidi WC, Carpenter CL, & Idle JR (1995) Polymorphism of glutathione S-transferase M1 and lung cancer risk among African-Americans and Caucasians in Los Angeles county, California. J Natl Cancer Inst, **87**: 1246-1253.

London SJ, Daly AK, Leathart JBS, Navidi WC, & Idle JR (1996) Lung cancer risk in relation to the CYP2C9*1/CYP2C9*2 genetic polymorphism among African-Americans and Caucasians in Los Angeles County, California. Pharmacogenetics, **6**: 527-533.

London SJ, Lehman TA, & Taylor JA (1997a) Myeloperoxidase genetic polymorphism and lung cancer risk. Cancer Res, **57**: 5001-5003.

London SJ, Sullivan-Klose T, Daly AK, & Idle JR (1997b) Lung cancer risk in relation to the CYP2C9 genetic polymorphism among Caucasians in Los Angeles County. Pharmacogenetics, **7**: 401-404.

London SJ, Idle JR, & Daly AK (1999) Genetic variation of CYP2A6, smoking, and risk of cancer. Lancet, **353**: 898-899.

Lucas D, Menez C, Girre C, Berthou F, Bodenez P, Joannet I, Hispard E, Bardou L-G, & Menez J-F (1995) Cytochrome P450 2E1 genotype and chlorzoxazone metabolism in healthy and alcoholic Caucasian subjects. Pharmacogenetics, **5**: 298-304.

MacKenzie PI, Rodbourn L, & Iyanagi T (1993) Glucuronidation of carcinogen metabolites by complementary DNA-expressed uridine 5'-diphosphate glucuronocyltransferases. Cancer Res, **53**: 1529-1533.

Mallet WG, Mosebrook DR, & Trush MA (1991) Activation of (+-)-trans-7,8-dihydroxy-7,8-dihydobenzo[a]pyrene to diolepoxides by human polymorpho-nuclear leucocytes or myeloperoxidase. Carcinogenesis, **12**: 521-524.

Marshall RS, Paterson MC, & Rauth AM (1991) DT-diaphorase activity and mitomycin C sensitivity in non-transformed cell strains derived from members of a cancer-prone family. Carcinogenesis, **12**: 1175-1180.

Martínez C, Agúndez JAG, Olivera M, Martín R, Ladero JM, & Benítez J (1995) Lung cancer and mutations at the polymorphic NAT2 gene locus. Pharmacogenetics, **5**: 207-214.

Mashimo M, Suzuki T, Abe M, & Deguchi T (1992) Molecular genotyping of *N*-acetylation polymorphism to predict phenotype. Hum Genet, **90**: 139-142.

Masimirembwa CM, Johansson I, Hasler JA, & Ingelman-Sundberg M (1993) Genetic polymorphism of cytochrome P450 CYP2D6 in a Zimbabwean population. Pharmacogenetics, **3**: 275-280.

Matthias C, Bockmühl U, Jahnke V, Jones PW, Hayes JD, Alldersea J, Gilford J, Bailey L, Bath J, Worrall SF, Hand P, Fryer AA, & Strange R (1998) Polymorphism in cytochrome P450 CYP2D6, CYP1A1, CYP2E1 and glutathione *S*-transferase, GSTM1, GSTM3, GSTT1 and susceptibility to tobacco-related cancers: studies in upper aerodigestive tract cancers.. Pharmacogenetics, **8**: 91-100.

Matthias C, Bockmühl U, Jahnke V, Harries L, Wolf CR, Jones PW, Alldersea J, Worrall SF, Hand P, Fryer AA, & Strange R (1998) The glutathione *S*-transferase GSTP1 polymorphism: effects on susceptibility to oral/ pharyngeal and laryngeal carcinomas. Pharmacogenetics, **8**: 1-6.

McBride OW, Umeno M, Gelboin HV, & Gonzalez FJ (1987) A Taq I polymorphism in the human P450IIE1 gene on chromosome 10 (CYP2E). Nucl Acid Res, **15**: 10071.

McGlynn KA, Rosvold EA, Lustbader ED, Hu Y, Clapper ML, Zhou T, Wild CP, Xia XL, Baffoe-Bonnie A, Ofori Adjei D, Chen G-C, London WT, Shen F-M, & Buetow KH (1995) Susceptibility to hepatocellular carcinoma is associated with genetic variation in the enzymatic detoxification of aflatoxin B1. Proc Natl Acad Sci, **92**: 2384-2387.

McLellan RA, Oscarson M, Seidegård J, Evans DAP, & Ingelman-Sundberg M (1997) Frequent occurrence of CYP2D6 gene duplication in Saudi Arabians. Pharmacogenetics, **7**: 187-191.

McLellan RA, Oscarson M, Hidestrand M, Leidvik B, Jonsson E, Otter C, & Ingelman-Sundberg M (2000) Characterization and functional analysis of two common human cytochrome P450 1B1 variants. Arch Biochem Biophys, **378**: 175-181.

McWilliams JE, Sanderson BJ, Harris EL, Richert-Boe KE, & Henner WD (1995) Glutathione S-transferase M1 (GSTM1) deficiency and lung cancer risk. Cancer Epidemiol Biomarkers Prev, **4**(6): 589-594.

Meisma R, Fernandez-Salguero P, van Kuilenburg ABP, van Gennip AH, & Gonzalez FJ (1995) Human polymorphism in drug metabolism: mutation in the dihydropyrimidine dehydrogenase gene results in exon skipping and thymine uracilurea. DNA Cell Biol, **14**: 1-6.

Messina ES, Tyndale RF, & Sellers EM (1997) A major role for CYP2A6 in nicotine C-oxidation by human liver microsomes. J Pharmacol Exp Ther, **282**: 1608-1614.

Meyer UA, Skoda RC, & Zanger UM (1990) The general polymorphism of debrisoquine/sparteine metabolism-molecular mechanisms. Pharmacol Ther, **46**: 297-308.

Meyer UA (1994) Pharmacogenetics: The slow, the rapid, and the ultrarapid. Proc Natl Acad Sci USA, **91**:1983-1984.

Michejda CJ, Kroeger, & Koepke MB (1994) Carcinogen activation by sulfate conjugate formation. Adv Pharmacol, **27**: 331-363.

Millikan RC, Pittman GS, Newman B, Tse C-KJ, Selmin O, Rockhill B, Savitz DS, Moorman PG, & Bell DA (1998) Cigarette smoking, N-acetyltransferases 1 and 2, and breast cancer risk. Cancer Epidemiol Biomarkers Prev, **7**: 371-378.

Minchin RF, Reeves PT, Teitel CH, McManus ME, Mojarrabbi B, Ilett KF, & Kadlubar FF (1992) N- and O-acetylation of aromatic and heterocyclic amine carcinogens by human monomorphic and polymorphic acetyltransferases expressed in COS-1 cells. Biochem Biophys Res Commun, **185**: 839-844.

Miyamoto M, Umetsu Y, Dosaka-Akita H, Sawamura Y, Yokota J, Kunitoh H, Nemoto N, Sato K, Ariyoshi, & Kamataki T (1999) CYP2A6 Gene deletion reduces susceptibility to lung cancer. Biochen Biophys Res Commun, **261**: 658-660.

Mrozikiewicz PM, Drakoulis N, & Roots I (1994) Polymorphic arylamine N-acetyltransferase (NAT2) genes in children with insulin-dependent diabetes mellitus. Clin Pharm Ther, **56**: 626-634.

Nebert DW (1989) The Ah locus: genetic differences in toxicity, cancer, mutation and birth defects. CRC Crit Rev Toxicol, **20**: 153-174.

Nebert DW, McKinnon RA, & Puga A (1996) Human drug-metabolizing enzyme polymorphisms: effects on risk of toxicity and cancer. DNA Cell Biol, **15**: 273-280.

Nelson DR, Koymans L, Kamataki T, Stegeman JJ, Feyereisen R, Waxman DJ, Waterman MR, Gotoh O, Coon MJ, Astabrook RW, Gunsalus IC, & Nebert DE (1996) P450 superfamily: update on new sequences, gene mapping, accession numbers and nomenclature. Pharmacogenetics, **6**: 1-42.

Norppa H, Hirvonen A, Järventaus H, Uusküla M, Tasa M, Ojajärvi A, & Sorsa M (1995) Role of GSTM1 and GSTT1 genotypes in determining individual sensitivity to sister chromatid exchange induction by diepoxybutane in cultured human lymphocytes. Carcinogenesis, **16**: 1261-1264.

Nunoya K, Yokoi T, Kimura K, Inoue K, Kodama T, Funayama M, Nagashima K, Funae Y, Green C, Kinoshita M, & Kamataki T (1998) A new deleted allele in the human cytochrome P450 2A6 (CYP2A6) gene found in individuals showing poor metabolic capacity to coumarin and (+)-cis-3,5-dimethyl-2-(3-pyridyl) thiazolidin-4-one hydrochloride (SM-12502). Pharmacogenetics, **8**: 239-249.

Oesch F (1973) Mammalian epoxide hydrolases: Inducible enzymes catalyzing the inactivation of carcinogenic and cytotoxic metabolites derived from aromatic and olefinic compounds. Xenobiotica, **3**: 305-340.

Oesch F, Glatt H, & Schimassmann H (1977) The apparent ubiquity of epoxide hydrolase in rat organs. Biochem Pharmacol, **26**: 603-607.

Omiecinski CJ, Aicher L, Holubkov R, & Checkoway H (1993) Human peripheral lymphocytes as indicators of microsomal epoxide hydrolase activity in liver and lung. Pharmacogenetics, **3**: 150-158.

Oscarson M, Gullstén H, Rautio A, Bernal ML, Sinues B, Dahl M-L, Stengård JH, Pelkonen O, Raunio H, & Ingelman-Sundberg M (1998) Genotyping of human cytochrome P450 2A6 (CYP2A6), a nicotine C-oxidase. FEBS Lett, **438**: 201-205.

Oscarson M, McLellan RA, Gullsten H, Agundez JAG, Benitez J, Rautio A, Raunio H, Pelkonen O, & Ingelman-Sundberg M (1999a) Identification and characterisation of novel polymorphisms in the CYP2A locus: implications for nicotine metabolism. FEBS Letters, **460**: 321-327.

Oscarson M, McLellan RA, Gullstén H, Yue Q-Y, Lang MA, Bernal ML, Sinues B, Hirvonen A, Raunio H, Pelkonen O, & Ingelman-Sundberg M (1999b) Characterisation and PCR-based detection of a CYP2A6 gene deletion found at a high frequency in a Chinese population. FEBS Lett, **448**: 105-110.

O'Shea D, Davis SN, & Kim RB (1994) Effect of fasting and obesity in humans on the 6-hydroxylation of chlorzoxazone: a putative probe of CYP2E1 activity. Clin Pharmacol Ther, **56**: 359-367.

Otto S, Bhattacharyya KK, & Jefcoate CR (1992) Polycyclic aromatic hydrocarbon in rat adrenal, ovary, and testis microsomes is catalyzed by the same novel cytochrome P450 (P450RAP), Endocrinology, **131**: 3067-3076.

Owens IS & Ritter JK (1995) Gene structure at the human UGT1 locus creates diversity in isozyme structure, substrate specificity, and regulation. Pror Nucleic Acid Res Mol Biol, **51**: 205-338.

Ozawa S, McDaniel LP, Tang Y-M, Schoket B, Vincze I, Kostic S, & Kadlubar FF (1997) CYP2C9 and GSTP1 genetic polymorphisms in patients with smoking-related lung cancer. Proc Am Ass Cancer Res, **38**: 212P.

Pelkonen O & Breimer DD (1994) Role of environmental factors in the pharmacokinetics of drugs - considerations with respect to animal models. In: Welling PG & Balant LP (eds) Handbook of experimental pharmacology. Basel, Karger, pp 289-332.

Pelkonen O & Raunio H (1995) Individual expression of carcinogen-metabolizing enzymes: cytochrome P4502A. J Occup Environ Med, **37**: 19-24.

Pelkonen O, Raunio H, Rautio A, Mäenpää J, & Lang MA (1993) Coumarin 7-hydroxylase: characteristics and regulation in mouse and man. J Irish Coll Phys Surg, **22**: 24-28.

Pelkonen O, Rautio A, & Raunio H (1995) Specificity and applicability of probes for drug metabolizing enzymes. In: Alvan G, Balant LP, Bechtel PR, Boobis AR, Gram LF, Paintaud G, & Pithan K (eds) European cooperation in the field of scientific and technical research - COST B1 conference on the variability and specificity in drug metabolism. Luxembourg, European Commission, pp 147-158.

Pelkonen O, Rautio A, Raunio H, & Pasanen M (2000) CYP2A6: a human coumarin 7-hydroxylase. Toxicology, **144**: 139-147.

Pemble S, Schroeder KR, Spencer SR, Meyer DJ, Hallier E, Bolt HM, Ketterer B, & Taylor JB (1994) Human glutathione S-transferase theta (GSTT1): cDNA cloning and the characterization of a genetic polymorphism. Biochem J, **300**: 271-276.

Persson I, Johansson I, & Ingelman-Sundberg M (1997) In vitro kinetics of two human CYP1A1 variant enzymes suggested to be associated with interindividual differences in cancer susceptibility. Biochem Biophys Res Commun, **231**: 227-230.

Peter R, Bocker R, Beaune PH, Iwasaki M, Guengerich FP, & Yang CS (1990) Hydroxylation of chlorzoxazone as a specific probe for human liver cytochrome P-450IIE1. Chem Res Toxicol, **3**: 566-573.

194

Philips IR, Dolphin CT, Clair P, Hadley MR, Hutt AJ, McCombie RR, Smith RL, & Shephard EA (1995) The molecular biology of the flavin-containing monooxygenases of man. Chem Biol Interact, **96**: 17-32.

Pianezza ML, Sellers EM, & Tyndale RF (1998) Nicotine metabolism defect reduces smoking. Nature, **393**: 750.

Probst-Hensch NM, Haile RW, Li DS, Sakamoto GT, Louie AD, Lin BK, Frankl HD, Lee ER, & Lin HJ (1996) Lack of association between the polyadenylation polymorphism in the NAT1 (acetyltransferase 1) gene and colorectal adenomas. Carcinogenesis, **17**: 2125-2129.

Raaka S, Hasset C, & Omiecinski CJ (1998) Human microsomal epoxide hydrolase: 5'-flanking region genetic polymorphism. Carcinogenesis, **19**: 387-393.

Raunio H, Husgafvel-Pursiainen K, Anttila S, Hietanen E, Hirvonen A, & Pelkonen O (1995a) Diagnosis of polymorphisms in carcinogen-activating and inactivating enzymes and cancer susceptibility-review. Gene, **159**: 113-121.

Raunio H, Pasanen M, Mäenpää J, Hakkola J, & Pelkonen O (1995b) Expression of extrahepatic cytochrome P450 in humans. In: Pacifici GM, & Fracchia GN (eds) Advances in drug metabolism in man. Luxembourg, European Commission, Office for Official Publications of the European Communities, pp 234-287.

Rautio A, Kraul H, Kojo A, Salmela E, & Pelkonen O (1992) Interindividual variability of coumarin 7-hydroxylation in healthy volunteers. Pharmacogenetics **2**: 227-233.

Risch A, Wallace DMA, Bathers S, & Sim E (1995) Slow *N*-acetylation genotype is a susceptibility factor in occupational and smoking related bladder cancer. Hum Mol Gen, **4**: 231-236.

Rebbeck TR (1997) Molecular epidemiology of the human glutathione *S*-transferase genotypes GSTM1 and GSTT1 in cancer susceptibility. Cancer Epid Biomarkers & Prevention, **6**: 733-743.

Rebbeck TR, Jaffe JM, Walker AH, Wein AJ, & Malkowicz SB (1998) Modification of clinical presentation of prostate tumors by a novel genetic variant in CYP3A4. J Natl Cancer Inst, **90**: 1225-1229.

Rettie AE, Wienkers LC, Gonzalez FJ, Trager WF, & Korzekwa KR (1994) Impaired (S)-warfarin metabolism catalysed by the $R^{144}C$ allelic variant of CYP2C9. Pharmacogenetics, **4**: 39-42.

Rojas M, Camus AM, Alexandrov K, Husgafvel-Pursiainen K, Anttila S, Vainio H, & Bartsch H (1992) Stereoselective metabolism of (-)-benzo(a)pyrene-7,8-diol by human lung microsomes and peripheral blood lymphocytes: effects of smoking. Carcinogenesis, **13**: 929-933.

Rostami-Hodjegan A, Lenard MS, Woods HE, & Tucker GT (1998) Meta-analysis of studies of the CYP2D6 polymorphism in relation to lung cancer and Parkinson's disease. Pharmacogenetics, **8**: 227-238.

Rosvold EA, McGlynn KA, Lustbader ED, & Buetow KH (1995) Identification of an NAD(P)H:quinone oxidoreductase polymorphism and its association with lung cancer and smoking. Pharmacogenetics, **5**: 199-206.

Rothman N, Hayes RB, Bi W, Caporaso N, Broly F, Woosley RL, Yin S, Feng P, You X, & Meyer UA (1993) Correlation between *N*-acetyltransferase activity and NAT2 genotype in Chinese males. Pharmacogenetics, **3**: 250-255.

Ryberg D, Skaug V, Hewer A, Phillips DH, Harries LW, Wolf CR, Øgreid D, Ulvik A, Vu P, & Haugen A (1997) Genotypes of glutathione transferase M1 and P1 and their significance for lung DNA adduct levels and cancer risk. Carcinogenesis, **18**: 1285-1289.

Saarikoski S, Voho A, Reinikainen M, Anttila S, Karjalainen A, Malaveille C, Vainio H, Husgafvel-Pursiainen K, & Hirvonen A (1998) Combined effect of polymorphic GST genes on individual susceptibility to lung cancer. Int J Cancer (1998).

Sachse C, Brockmöller J, Bauer S, & Roots I (1997) Cytochrome P450 2D6 variants in a Caucasian population: allele frequencies and phenotypic consequences. Am J Hum Genet, **60**: 284-295.

Sata F, Sapone A, Elizondo G, Stocker P, Miller VP, Zheng W, Raunio H, Crespi CL, & Gonzalez FJ (2000) CYP3A4 allelic variants with amino acid substitutions in exons 7 and 12: evidence for an allelic variant with altered catalytic activity. Clin Pharmacol Ther, **67**: 48-56.

Schmekel B, Karlsson SE, Linden M, Sundström C, Tenge H, & Venge P (1990a) Myeloperoxidase in human lung lavage. I. A marker of local neutrophil activity. Inflammation, **14**: 447-454.

Schmekel B, Hornblad Y, Linden M, Sundström C, & Venge P (1990b) Myeloperoxidase in human lung lavage. II. Internalization of myeloperoxidase by alveolar macrophages. Inflammation, **14**: 455-461.

Schröder KR, Hallier E, Peter H, & Bolt HM (1992) Dissociation of a new glutathione *S*-transferase activity in human erythrocytes. Biochem Pharmacol, **43**: 1671-1674.

Schulz WA, Krummeck A, Rösinger I, Eickelmann P, Neuhaus C, Ebert T, Schmitz-Dräger BJ, & Sies H (1997) Increased frequency of a null-allele for NAD(P)H:quinone oxidoreductase in patients with urological malignancies. Pharmacogenetics, **7**: 235-239.

Seidegård J & Ekström G (1997) The role of human glutathione transferases and epoxide hydrolases in the metabolism of xenobiotics. Environ Health Perspect, **105**: 791-799.

Seidegård J, Vorachek WR, Pero RW, & Pearson WR (1988) Hereditary differences in the expression of the human glutathione transferase active on trans-stilbene oxide are due to a gene deletion. Proc Natl Acad Sci USA, **85**: 7293-7297.

Shields PG, Sugimura H, Caporaso NE, Petruzzelli SF, Bowman ED, Trump BF, Weston A, & Harris CC (1992) Polycyclic aromatic hydrocarbons-DNA adducts and the CYP1A1 restriction fragment length polymorphism. Environ Health Perspect, **98**: 191-194.

Shimada T, Yamasaki H, Mimura M, Inui Y, & Guengerich FP (1994) Interindividual variations in human liver cytochrome P450 enzymes involved in the oxidation of drugs, carcinogens and toxic chemicals: studies with liver microsomes of 30 Japanese and 30 Caucasians. J Pharmacol Exp Ther, **270**: 414-423.

Shimada T, Hayes CL, Yamazaki H, Amin S, Hecht SS, Guengerich FP, & Sutter TR (1996a) Activation of chemically diverse procarcinogens by human cytochrome P-450 1B1, Cancer Res, **56**: 2979-2984.

Shimada T, Yamazaki H, & Guengerich FP (1996b) Ethnic-related differences in coumarin 7-hydroxylation activity catalyzed by cytochrome P2402A6 in liver microsomes of Japanese and Caucasians populations. Xenobiotica, **26**: 395-403.

Shou M, Krausz KW, Gonzalez FJ, & Gelboin HV (1996) Metabolic activation of the potent carcinogen dibenzo(a)pyrene by human recombinant cytochromes P450, lung and liver microsomes. Carcinogenesis, **17**: 2429-2433.

Sims P, Grover PL, Swaisland A, Pal K, & Hewer A (1974) Metabolic activation of benzo(a)pyrene proceeds by a diol epoxide. Nature, **252**: 326-328.

Smith CAD & Harrison DJ (1997) Association between polymorphism in gene for microsomal epoxide hydrolase and susceptibility to emphysema. Lancet, **350**: 630-633.

Smith G, Stanley LA, Sim E, Strange R, & Wolf CR (1995) Metabolic polymorphisms and cancer susceptibility. Cancer Surv, **25**: 27-65.

Spink DC, Hayes CL, Young NR, Christou M, Sutter TR, Jefcoate CR, & Gierthy JF (1994) The effect of 2,3,7,8-tetrachloro-dibenzo-p-dioxin on estrogen metabolism in MCF-7 breast cancer cells: evidence for induction of a novel 17b-estradiol-4-hydroxylase. J Steroid Biochem Mol Biol, **51**: 251-258.

Stoilov I, Akarsu AN, & Sarfarazi M (1997) Identification of three different truncating mutations in cytochrome P4501B1 (CYP1B1) as the principal cause of primary congenital glaucoma (Buphthalmos) in families linked to the GLC3A locus on chromosome 2p21. Human Mol Genet, **6**: 641-647.

Sugimura T (1992) Multistep carcinogenesis: a 1992 perspective. Science, **258**: 603–607.

Sugimura H, Wakai K, Genka K, Nagura K, Igarashi H, Nagayama K, Ohkawa A, Baba S, Morris BJ, Tsugane S, Ohno Y, Gao CM, Li ZY, Takezaki T, Tajima K, & Iwamasa T (1998) Association of Ile462Val (exon 7) polymorphism of cytochrome P450 1A1 with lung cancer in the Asian population: further evidence from a case-control study in Okinawa. Cancer Epidemiol Biomarkers Prev, **7**: 413-417.

Sullivan-Klose TH, Ghanayem BI, Bell DA, Zhang Z-Y, Kaminsky LS, Shenfield GM, Miners JO, Birkett DJ, & Goldstein JA (1996) The role of the CYP2C9-Leu359 allelic variant in the tolbutamide polymorphism. Pharmacogenetics, **6**: 341-349.

Sutter TR, Guzman K, Dold KM, & Greenlee WF (1991) Targets for dioxin: genes for plasminogen activator inhibitor-2 and interleukin-1β. Science, **254**: 415-418.

Sutter TR, Tang YM, Hayes CL, Wo Y-Y P, Jabs EW, Li X, Yin H, Cody CW, & Greenlee WF (1994) Complete cDNA sequence of a human dioxin-inducible mRNA identifies a new gene subfamily of cytochrome P450 that maps to chromosome 2*. J Biol Chem, **269**: 13092-13099.

Swanson HI & Bradfield CA (1993) The AH-receptor: genetics, structure and function. Pharmacogenetics, **3**: 213-230.

Taylor JA, Umbach D, Stephens E, Castranio T, Paulson D, Robertson C, Mohler JL, & Bell DA (1998) The role of N-acetylation polymorphism in smoking associated bladder cancer: Evidence of a gene-gene-exposure three-way interaction. Cancer Res, **58**: 3603-3610.

Tefre T, Ryberg D, Haugen A, Nebert DW, Skaug V, Brøgger A, & Børresen AL (1991) Human CYP1A1 (cytochrome P_1450) gene: lack of association between the MspI restriction fragment length polymorphism and incidence of lung cancer in a Norwegian population. Pharmacogenetics, **1**: 20-25.

To-Figueras J, Gené M, Gómez-Catalán J, Galán MC, Fuentes M, Ramón JM, Rodamilans M, Huguet E, & Corbella J (1997) Glutathione S-transferase M1 (GSTM1) and T1 (GSTT1) polymorphisms and lung cancer risk among Northwestern Mediterraneans. Carcinogenesis, **18**: 1529-1533.

Traver RD, Horikoshi T, Danenberg PV, Ross D, & Gibson NW (1992) NAD(P)H:quinone oxidoreductase gene expression in human colon carcinoma cells: characterization of mutation which modulates DT-diaphorase activity and mitomycin sensitivity. Cancer Res, **52**: 797-802.

Tsuneoka Y, Fukushima K, Matsuo Y, Ichikawa Y, & Watanabe Y (1996) Genotype analysis of the CYP2C19 gene in the Japanese population. Life Sci, **59**: 1711-1715.

Tucker GT, Rostami-Hodjegan A, Nurminen S, & Jackson PR (1995) Phenotyping populations: Pharmacokinetic and statistical issues. In: Alvan G, Balant LP, Bechtel PR, Boobis AR, Gram LF, Paintaud G, & Pithan K (eds) European cooperation in the field of scientific and technical research - COST B1 conference on the variability and specificity in drug metabolism. Luxembourg, European Commission, pp 191-203.

Turesky RJ, Lang NP, Butler MA, Teitel CH, & Kadlubar FF (1991) Metabolic activation of carcinogenic heterocyclic aromatic amines by human liver and colon. Carcinogenesis, **12**: 1839-1845.

Vatsis KP, Weber WW, Bell DA, Dupret J-M, Evans DAP, Grant DM, Hein DW, Lin HJ, Meyer UA, Relling MV, Sim E, Suzuki T, & Yamazoe Y (1995) Nomenclature for N-acetyltransferases, Pharmacogenetics, **5**: 1-17.

Uematsu F, Kikuchi H, Motomiya M, Abe T, Ishioka C, Kanamaru R, Sagami I, & Watanabe M (1992) Human cytochrome P450IIE1 gene: DraI polymorphism and susceptibility to cancer. Tohoku J Exp Med, **168**: 113-117.

Uematsu F, Ikawa S, Sagami I, Kanamaru R, Abe T, Satoh K, Motomiya M, & Watanabe M (1994) Restriction fragment length polymorphism of the human CYP2E1 (cytochrome P450IIE1) gene and susceptibility to lung cancer: possible relevance to low smoking exposure. Pharmacogenetics, **4**: 58-63.

Vatsis KP & Weber WW (1993) Structural heterogeneity of Caucasian N-acetyltransferase at the NAT1 gene locus. Arch Biochem Biophys, **301**: 71-76.

Vatsis KP, Weber WW, Bell DA, Dupret J-M, Evans DAP, Grant DM, Hein DW, Lin HJ, Meyer UA, Relling MV, Sim E, Suzuki T, & Yamazoe Y (1995) Nomenclature for N-acetyltransferases. Pharmacogenetics, **5**: 1-17.

Vessel ES, Seaton TD, & A-Rahim YI (1995) Studies on interindividual variations of CYP2E1 using chlorzoxazone as an *in vivo* probe. Pharmacogenetics, **5**: 53-57.

Vineis P, Bartsch H, Caporaso N, Harrington AM, Kadlubar FF, Landi MT, Malaveille C, Shields PG, Skipper P, Talaska G, & Tannenbaum SR (1994) Genetically based N-acetyltransferase metabolic polymorphism and low-level environmental exposure to carcinogens. Nature, **369**: 154-156.

Vineis P, Malats N, Lang M, d'Errico A, Caporaso N, Cuzick J, & Boffetta P (1999) IARC Scientific Publications No. 148. Metabolic Polymorphisms and Susceptibility to Cancer. Lyon, International Agency for Research on Cancer.

Vogelstein B & Kinzler KW (1993) The multistep nature of cancer. Trends Genet, **9**: 138-141.

Watanabe J, Hayashi S, & Kawajiri K (1994) Different regulation and expression of the human CYP2E1 gene due to the RsaI polymorphisms in the 5'-flanking region. J Biochem (Tokyo), **116**: 321-326.

Watanabe J, Shimada T, Gillam EM, Ikuta T, Suemasu K, Higashi Y, Gotoh O, & Kawajiri K (2000) Association of CYP1B1 genetic polymorphism with incidence to breast and lung cancer. Pharmacogenetics, **10**: 25-33.

Weber WW & Hein DW (1985) *N*-acetylation pharmacogenetics. Pharmacol Rev, **37**: 25-79.

Weber WW, Mattano SS, & Levy GN (1988) Acetylator pharmacogenetics and aromatic amine-induced cancer. In: Carcinogenic and mutagenic responses to aromatic amines and nitroarenes. King CM (ed.) New York, Elsevier.

Wedlund PJ, Kimura S, Gonzalez FJ, & Nebert DW (1994) 1462V mutation in the human CYP1A1 gene: lack of correlation with either the Msp I 1.9 kb (M2) allele of CYP1A1 inducibility in a three-generation family of East Mediterranean descent. Pharmacogenetics, **4**: 21-26.

Wei X, McLeod HL, McMurrough J, Gonzalez FJ, & Fernandez-Salguero P (1996) Molecular basis of the human dihydropyrimidine dehydrogenase deficiency and 5-fluorouracil toxicity. J Clin Invest, **98**: 610-615.

Weinshilboum RM (1992) Methylation pharmacogenetics: thiopurine methyl-transferase as a model system. Xenobiotica, **22**: 1055-1071.

Weinshilboum RM, Otterness DM, Aksoy LA, Wood TC, Her C, & Raffogianis RB (1997) Sulfotransferase molecular biology. FASEB J, **11**: 3-14.

Westlind A, Lofberg L, Tindberg N, Andersson TB, & Ingelman-Sundberg M (1999) Interindividual differences in hepatic expression of CYP3A4: relationship to genetic polymorphism in the 5'-upstream regulatory region. Biochem Biophys Res Commun, **259**: 201-205.

Wiencke JK, Pemble S, Ketterer B, & Kelsey KT (1995) Gene deletion of glutathione transferase theta 1: Correlation with induced genetic damage and potential role in endogeneous mutagenesis. Cancer Epid Biomarkers Prev, **4**: 253-260.

Wild D, Fesrs W, Michel S, Lord HL, & Josephy PD (1995) Metabolic activation of heterocyclic aromatic amines catalyzed by human arylamine *N*-acetyltransferase isozymes (NAT1 and NAT2) expressed in *Salmonella typhimurium*. Carcinogenesis, **16**: 643-648.

Wormhoudt LW, Commandeur JNM, & Vermeulen NPE (1999) Genetic polymorphisms of human *N*-acetyltransferase, cytochrome P450, glutathione-*S*-transferase, and epoxide hydrolase enzymes: relevance to xenobiotic metabolism and toxicity. Crit rev Toxicol, **29**: 59-124.

Wrighton SA & Stevens JC (1992) The human hepatic cytochromes P450 involved in drug metabolism. Crit Rev Toxicol, **22**: 1-21.

Xu X, Kelsey KT, Wiencke JK, Wain JC, & Christiani DC (1996) Cytochrome P450 CYP1A1 MspI polymorphism and lung cancer susceptibility. Cancer Epidemiol Biomarkers Prev, **5**: 687-692.

Yanagawa Y, Sawada M, Deguchi T, Gonzalez FJ, & Kamataki T (1994) Stable expression of human CYP1A2 and N-acetyltransferases in Chinese hamster CHL cells: mutagenic activation of 2-amino-3-methylimidazo[4,5-f]quinoline and 2-amino-3,8-dimethylimidazo [4,5]-f-quino-xaline. Cancer Res, **54**: 3422-3427.

Yengi L, Inskip A, Gilford J, Alldersea J, Bailey L, Smith A, Lear JT, Heagerty AH, Bowers B, Hand P, Hayes JD, Jones PW, Strange RC, & Fryer AA (1996) Polymorphism at the glutathione S-transferase locus GSTM3: interactions with cytochrome P450 and glutathione S-transferase genotypes as risk factors for multiple cutaneous basal cell carcinoma. Cancer Res, **56**: 1974-1977.

Yokoi T, Sawada M, & Kamataki T (1995) Polymorphic drug metabolism: studies with recombinant Chinese hamster cells and analyses in human populations. Pharmacogenetics, **5**: S65-S69.

Yuspa SH, Dlugosz AA, Cheng CK, Denning MF, Tannenbaum T, Glick AB, & Weinberg WC (1994) Role of oncogenes and tumor suppressor genes in multistage carcinogenesis. J Invest Dermatol, **103**: 90S-95S.

Zenser TV, Lakshmi VM, Rustan TD, Doll MA, Deitz AC, Davis BB, & Hein DW (1996) Human N-acetylation of benzidine: role of NAT1 and NAT2. Cancer Res, **56**: 3941-3947.

Zhang Z-Y, Fasco MJ, Huang L, Guengerich FP, & Kaminsky LS (1996) Characterization of purified human recombinant cytochrome P4501A1-Ile462 and -Val462: assessment of a role for the rare allele in carcinogenesis. Cancer Res, **56**: 3926-3933.

Zimniak P, Nanduri B, Pilula S, Bandorowicz-Pikula J, Singhal S, Srivastava SK, Awasthi S, & Awasrhi JC (1994) Naturally occurring human glutathione S-transferase GSTP1.1 isoforms with isoleucine and valine at position 104 differ in enzymatic properties. Eur J Biochem, **224**: 893-899.

VALIDATION OF BIOMARKERS FOR ENVIRONMENTAL HEALTH RESEARCH AND RISK ASSESSMENT

Paul A. Schulte

Education and Information Division, National Institute for Occupational Safety and Health, Cincinnati, Ohio 45226, USA

CONTENTS

IV.1. INTRODUCTION

The term "biological markers" or biomarkers" refers to indicators of events in biological systems or samples (NRC, 1987, 1989a). It is useful to classify biomarkers into three types: exposure, effect and susceptibility. Many of the characteristics and issues pertaining to each type of biomarkers are different, and the three types of biomarkers should be considered separately.

A biological marker of exposure is an exogenous substance or its metabolite or the product of a xenobiotic agent and some target molecule or cell that is measured in a compartment within an organism. A biological marker of effect is a measure of molecular, biochemical or physiological change or other alteration within an organism that can be recognized as an established or potential health impairment or disease. A biomarker of susceptibility is an indicator of an inherited or acquired limitation of an organism's ability to respond to the challenge of exposure to a xenobiotic substance (NRC, 1989a).

Validity refers to a range of characteristics that is the best approximation of the truth or falsehood of a biomarker. It is a sense of degree rather than an all-or-none state. Three broad categories of validity can be distinguished: measurement validity, internal study validity and external validity. Measurement validity is the degree to which a biomarker indicates what it purports to indicate. Internal study validity is the degree to which inferences drawn from a study actually pertain to study subjects and are true. External validity is the extent to which findings of a study can be generalized to apply to other populations (Schulte & Perera, 1993).

The use of invalid biomarkers can lead to invalid inferences and generalizations and ultimately to erroneous risk assessments. Validation is a multistage process that involves laboratory-based and epidemiological assessments.

IV.2. STATUS OF BIOMARKERS IN ENVIRONMENTAL HEALTH EFFORTS

Despite stunning developments in the laboratory (e.g., Dennisenko et al., 1996) and a few exciting efforts in population studies (Galloway et al., 1986; Ross et al., 1992; Perera et al., 1992), most biomarkers (other than genotypes for single disease) have not been used to identify new causes of disease, serve as sentinels of early, reversible, or more treatable conditions, serve as effect modifiers of exposure-related health effects, or been used in risk assessments. On the other hand, much of the published literature on biomarkers has provided important and useful contributions to understanding mechanisms, particularly of carcinogens, or has identified genotypes indicative of risk for inherited diseases. Generally, the use of biomarkers (with the exception of markers of internal dose, e.g., blood lead or markers of exposures in infectious disease, or serum lipids in cardiovascular disease) for studying non-malignant environmentally induced diseases has been less than for malignant diseases. In short, the work to bring a biomarker from the laboratory to the field and develop it so that it can be useful as a dependent or independent variable in an epidemiological study, as an indicator in population surveillance, or to contribute to quantitative risk assessments has been lacking.

Although biomarkers have a long history in medicine and public health, the systematic development, validation and application of biomarkers is a relatively new field (Shugart et al., 1992). Ward & Henderson (1996) have identified four research needs that could develop this field. First, additional new biomarkers need to be developed or existing ones refined to fill gaps in the continuum of events from environmental exposure to clinical disease expression. Second, there is a need to better understand the relationships of specific biomarkers to the pathophysiological mechanisms of disease to estimate more accurately risk of disease. Third, biomarkers need to be better characterized with respect to their sensitivity, specificity and variability, and validated as predictors of adverse health effects. Fourth, there is a need to address the many societal impediments to the validation and practical use of biomarkers as public health tools. In each of these efforts, laboratory scientists and epidemiologists, clinicians, exposure assessors and statisticians need to be involved. In addressing societal impediments, an even broader range of disciplines, such as ethicists, lawyers, economists and insurers also needs to participate.

The optimal use of biomarkers will most likely occur if environmental health research is linked not only to epidemiological studies but to studies of laboratory animals and cell lines and to the assessment of biomarkers in ecological studies of plants and animals in the wild (Shugart et al., 1992; Anderson et al., 1994). Biomarkers can serve as a common element in studies of these different groups or materials. Thus, a biomarker identified in an exposed laboratory animal or cell line might also be seen in wild or laboratory animals or humans with similar exposures. A parallelogram type approach (Sobels, 1993; Sutter, 1995) can be used to assess the relationship between markers and risks in those groups (Fig. 15). The parallelogram approach is derived from the 1970s work of Sobels (1993) to extrapolate damage from animals to humans. Genetic damage that cannot be measured directly, such as in human germ cells, can be estimated by measuring the same kind of damage in both germ cells and somatic cells of the mouse. With data on the induction of mutations or chromosome aberrations in both germ cells and somatic cells of the mouse, it is possible to estimate germ cell mutation frequencies in humans on the basis of what can be measured by monitoring genetic damage in human somatic cells (Sobels, 1993). Sutter (1995) has modified this approach to include *in vitro* to *in vivo* extrapolation. In the parallelogram experimental approach to knowledge of mechanism, *in vitro* data is used to test the hypothesis that a specific mechanism of action is conserved among rodent and human species (Sutter, 1995). Biomarkers can be used to reduce high to low dose and species extrapolation-related uncertainties by providing information on common mechanisms and the development of mechanistically based mathematical models (Sexton et al., 1995). The incorporation of biomarkers of exposure and susceptibility in physiologically based pharmacokinetic (PBPK) models has allowed for interspecies comparisons and enabled the simulation of different enzyme activities among individuals (Fennell et al., 1996). Biomarkers also may serve as an alternative to the use of PBPK models for determining dose (Rohmberg, 1995). They are particularly useful when they are more easily or accurately measured than the actual exposure. Since biomarkers have extraordinary sensitivity, they may significantly extend the range of empirical characterization of dose and response in cases where they may be detected and measured at dose levels below those at which other effects are directly observable (Rohmberg, 1995; Ehrenberg et al., 1996). For example, adduct measurements of some alkylating agents may be used to indicate disease risks at levels too low to be detected by epidemiological means (Ehrenberg et al., 1996).

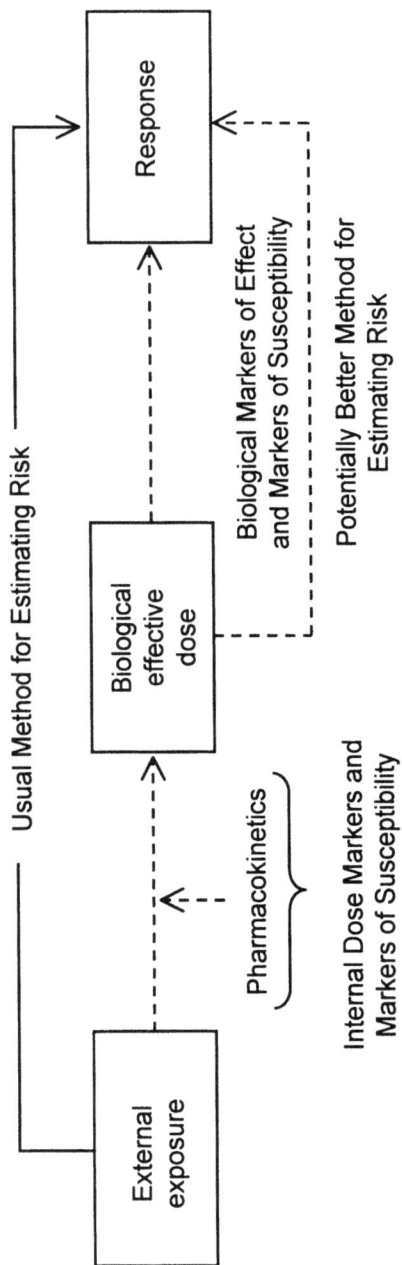

Fig. 15. Rationale for using biomarkers to assess risk
(Schute & Waters, 1999)

207

In risk assessment, valid biomarkers are those that not only represent events in mechanistic pathways, but also provide resolution of uncertainties that otherwise are addressed by default assumptions. Hattis & Silver (1993) describe the need for cross-cutting research that can validate relationships between scientific models, biomarkers and human risk. A major challenge in risk assessment is to demonstrate the biological plausibility and clinical significance of conclusions from epidemiological, lifetime animal, and short-term studies of chemicals thought to have potential adverse effects in human health and the environment. Biomarkers can address this challenge by linking the presence of a chemical in various environmental compartments to specific sites of action in target organs and to host responses (Perera, 1987; Omenn, 1995; Perera, 1996).

To have validity for risk assessment, there must be confidence that biomarkers that are assessed at high levels of exposure are relevant at lower levels (Perera, 1987; Perera et al., 1989; McClellan, 1995). Risk assessment requires integrating levels of information from multiple levels and sources (Hattis & Silver, 1993; McClellan, 1995). Thus, valid biomarkers need to be coupled with epidemiological and toxicological data to give a complete picture of risk. Characterizing risk involves conducting mechanistic research, identifying appropriate biomarkers and statistical models, and continuously evaluating the assumptions (default options) that are used when there is missing information or uncertainty. A valid biomarker for risk assessments will be one that helps resolve major rather than trivial uncertainties (McClellan, 1995).

IV.3. VALIDATION

The ultimate driving force for whether biomarkers will contribute to environmental health efforts is the validity of the markers. Validity is a complex characteristic that describes the extent to which a biomarker reflects a designated event in a biological system. Generally, these events are exposure, effects of exposure or disease, and susceptibility.

Validity has meaning according to discipline as well. To the laboratory scientist, validity often refers to the nature of the biomarker and the characteristic of the assay for the biomarker.

Thus, the sensitivity of the assay to detect a signal at a given concentration, and the ability of the signal to be specific for a particular event are indications of validity to the laboratory scientist. In addition, the scientist wants to know what factors might influence an assay.

The epidemiologist relies on the laboratory definition of validity as the cornerstone of population studies, but then needs to know how likely a person with a positive assay or test is to develop disease (or have been exposed) and how likely a person with a negative test is to be free of disease (or exposure). The epidemiologist also needs to know how feasible the marker is to use in human populations and the reliability of the assay under field conditions. Moreover, the epidemiologist needs to know how the frequency of the marker varies in different population subgroups defined by age, race, gender, pre-existing illness, diet and various behavioural factors. Only when validity at the laboratory and population level has been established is a biomarker ready for the full spectrum of environmental research and uses. As noted, most biomarkers have not had that level of validation. A broad effort is underway, but the products of this activity are not available yet.

Validation of candidate biomarkers is an empirical process that can be approached by producing several different but convergent lines of evidence. There is an extensive literature on criteria for validating biological markers (e.g., WHO, 1975; Gann, 1986; Lucier & Thompson, 1987; Hernberg & Aitio, 1987; Schulte, 1989; Schatzkin et al., 1990; Margetts, 1991; Schulte & Mazzuckelli, 1991; Stevens et al., 1991; Schulte & Perera, 1993; Schulte & Talaska, 1995). In general, these criteria include understanding the natural history, biological and temporal relevance, pharmacokinetics, background variability, dose-response, and confounding factors (Schulte & Talaska, 1995). Biomarker validity also depends on reliability of the assay to measure the biomarkers. These criteria allow for the assessment of whether a biomarker represents an event that is in a continuum between exposure and resultant disease, whether the biological specimen containing the biomarkers is appropriate, and whether the marker reflects the time period of concern. Finally, by assessing confounding and effect-modifying factors, it is possible to understand what other factors influence a biomarker or its assay.

The careful measurements of strong confounders and effect modifiers should be given as much attention as is given to measurement of the exposure and disease variables or biomarkers. Consideration should be given to mounting validation substudies to quantify measurement error in important covariates (Hatch & Thomas, 1993). Measurements of biological markers are the building blocks of research and risk assessment. If the measurements are invalid, the research and risk assessments are also likely to be invalid. Controlling measurement validity makes it possible partially to control study validity since measurement errors can produce biased estimates of regression coefficients used in statistical models of exposures and disease (Louis, 1988). Measures of association, such as the odd ratios, can be distorted, depending on the type of error and other characteristics, toward or away from the null hypotheses of no association between the biomarker and disease (or exposure).

As White (1997) notes, measurement errors for an individual can be defined as the difference between a person's measured biomarker (the biomarker "test") and the person's true biomarker. The true biomarker can be conceptualized as the underlying biomarker without laboratory or other sources of error, and if the measure can fluctuate over time, the true biomarker would be integrated over the time period of etiological interest. There are numerous sources of measurement error in biomarkers, some of which are shown in Table 15.

Validity in this context can be defined as the relation of the biomarker test (the potentially mismeasured biomarker) to the true biomarker in the population of interest. Measures of validity are parameters that describe the error in the population (White, 1997). Two measures of measurement error are used to describe the validity of an observed measurement compared with the true measurement (Armstrong et al., 1994). The first is systematic error or bias that would occur on average for subjects measured. The second is subject error, which is additional error that varies from subject to subject. The subject error is also known as precision or the measure of the variation of measurement error in the population. Precision can be assessed by a construct known as the validity coefficient. It ranges from 0 to 1 with the value 1 indicating that the observed measurement is a perfectly precise measure of the true measurement

Table 15. Examples of sources of measurement error in laboratory measures
in epidemiological studies

Errors in the laboratory method as a measure of the exposure of interest

Method may not measure all sources of the biological true exposure
interest
Method may measure other exposures that are not the true exposure
of interest
Methods may be influenced by subject characteristics (other than the
true exposure) that the researcher cannot manipulate, e.g., by the
disease under study, by other diseases

Errors or omissions in the protocol

Failure to specify the protocol in sufficient detail regarding timing and
method of specimen collection, specimen handling, storage, and
laboratory analytical procedures
Failure to include standardization of the instrument periodically
throughout the data collection

Errors due to biological variability within subjects

Short-term variability (hour to hour, day to day) in biological characteristics due to, for example, diurnal variation, time since last meal,
posture (sitting versus lying down)
Medium-term variability (month to month) due to, for example,
seasonal changes in diet
Long-term change (year to year) due to, for example, purposeful
dietary changes over time

Errors due to variation in execution of the protocol

Variations in method of specimen collection
Variations in specimen handling or preparation
Variations in length of specimen storage
Variations in specimen analysis between batches (different batches of
chemicals, different calibration of instrument)
Variation in technique between laboratory technicians
Random error within batch

White (1997)

211

(Armstrong et al., 1994). A validity study is defined here as one in which a sample of individuals are measured twice: once using the biomarker test of interest and once using a perfect or near perfect measure of the true biomarker (White, 1997). Then the measure of biomarker measurement error from the validity study can be applied to what is known about the association under study in the parent study to estimate the effects of biomarker error on the association of interest (White, 1997). While the impact of measurement error on exposure–disease associations has been studied extensively, the impact on estimates of interaction of two or more risk factors has been studied less thoroughly (Greenland, 1993). Assessment of interaction of multiple exposures, gene–environment, or gene–gene is an important issue in environmental epidemiology and all the more important with biomarkers depicting mechanistic events.

The demand on environmental epidemiology to evaluate increasingly subtle health risks requires more accurate estimation of the quantity and timing of a toxicant reaching target tissue (Kriebel, 1994). Kriebel (1994) has described a two-stage approach to derive estimates of dose from exposure data and then link them to epidemiological models estimating disease risk. Such an approach incorporates physiological processes into epidemiological modelling and is possibly more valid than approaches with less detail. Other approaches that may be useful to evaluate subtle health risks include: exploring the "genetic model" for explaining environmentally induced disease; using transgenic technologies; assessing the role of proteins as messengers and receptors; using computer technology to manipulate chemical structures; and conducting multisystem studies (Olden & Klein, 1995).

Ultimately, validation requires the use of epidemiological study designs to assess three relationships: exposure–dose; biological effects–disease; and susceptibility influencing an exposure–disease relationship. Studies that contribute to these types of validation and bridge the gap between laboratory experimentation and population–based epidemiology have been referred to as "transitional" studies (Hulka, 1991; Schulte et al., 1993; Rothman et al., 1995). They may be designed to evaluate exposures, health effects or susceptibility, and some may have the characteristics of pilot or developmental studies (Hulka & Margolin, 1992).

IV.4. VALIDATION OF EXPOSURE BIOMARKERS

Exposure → Internal → Biologically → Disease
                    dose   effective
                                     dose

A valid marker of exposure will not completely correspond to exogenous exposure since it may reflect various host factors (such as phase I or phase II enzymatic activity) and various routes and sources or exposure. Although some correspondence with exposure is important, the marker may actually be a better measure of exposure and hence not be strongly correlated with exposure (Hulka, 1991; Vine, 1996). The true test of the validity of an exposure biomarker is to determine if it is predictive of some health outcome or risk (Vine, 1996). However, if the goal is to have merely a useful measure of exposure that has better characteristics than some other measure of exposure, then comparison with that "other" measure is appropriate.

As the initial step in the validation process, it is important to know how the biomarkers of exposure correspond to measures of exogenous exposure. Critical in such research is the need to have an effective exposure assessment. This may require a combination of personal and environmental monitoring and questionnaires, record review and modelling to reconstruct exposure history. The approach also requires understanding of the toxico- and pharmacokinetics involved for the particular xenobiotics (Bernard, 1995). Related to this is the need to understand the natural history of the marker and utilize the information in the validation study. For example, in a study of hydroxyethyl haemoglobin adducts in workers exposed to ethylene oxide, we used the lifespan of the erythrocyte (approximately 4 months) as the time span in which to reconstruct exposure (Schulte et al., 1992). There is also a need to account for factors that might influence the appearance of a marker (Alessio et al., 1995). Reference data are generally required so that the level of the marker in unexposed populations can be assessed (Schulte et al., 1992; Grandjean et al., 1995). In the aforementioned study (Schulte et al., 1992), an exposure–response relationship between ethylene oxide and hydroxyethyl haemoglobin adducts was found at levels below the permissible exposure level when mean adduct values were adjusted for important covariates such as age, cigarette smoking and education (Schulte et al., 1992).

Biomarkers also may provide a useful indication of both exposure and host factors and thus be a better variable than simply exogenous exposure. The question to ask is whether the biomarker offers more information than an exposure assessment or questionnaire. A particular limitation is that many biomarkers of exposure reflect only relatively recent time (Pearce et al., 1995; Vineis & Porta, 1996). Thus, unless there is a constant pattern of exposure, most biomarkers of exposure will not be useful in epidemiological research of historic exposures. If the exposure biomarker is to be used to predict health risk, there is a need for validation studies. Verberk (1995) has provided guidelines for choosing between an external exposure (ExEx) and a biomarker of exposure (BmEx) to estimate health risk of a specific part of a population due to an environmental factor. These are listed in Table 16.

IV.5. VALIDATION OF THE RELATIONSHIP BETWEEN BIOLOGICAL EFFECTS AND DISEASE

Early biological $\rightarrow$ Altered structure $\rightarrow$ Disease
effects   and function

Most often, the term "validation" concerns the meaning of the biomarker with regard to disease. There are two approaches to the identification of disease from recognition of a biomarker: the "determinist" decision made by the individual physician, and the "probabilist" approach of the epidemiologist (Goyer & Rogan, 1986). The physician makes a diagnosis mostly through judgement but also by drawing on knowledge of mechanisms or pathophysiology involved. The epidemiologist on the other hand relates the prevalence of a biomarker of effect to the development of disease. This is largely a statistical approach.

The relationship of non-specific biological markers of effect to disease can be diluted or confounded from exposures other than those of interest. As Wilcosky notes: "If two different exposures, E_1 and E_2, cause the same marker response through independent

Table 16. Biomarkers of exposure (BmEx) versus parameters of
external exposure (ExEx)

1) Local effects (directly on airways, eyes or skin) usually require ExEx, whereas systemic effects call for BmEx.
2) Determination of the contribution of a specific, environmental source among multiple, e.g., non-environmental, sources of a substance calls for ExEx.
3) The availability of a reliable exposure–response relationship for the effect considered and of a health-based limit value; theoretically a relationship based on BmEx can be more reliable.
4) The possibility to determine reliably the exposure data that are needed, depending on the time and duration of the sampling with respect to the pattern of the external exposure, on the number of samples, and the toxicokinetic properties.
5) Inconvenient route of entry points to BmEx.
6) The presence of a group at risk due to intake-related behaviour or toxicokinetics calls for BmEx.
7) In the case of non-specificity of BmEx due to other substances, the increase or decrease of the effect predicting value should be evaluated.
8) Substantial probability of effects calls for BmEx.
9) Feasibility of sampling technique and reliability of the analysis
10) Acceptance by the public points to BmEx
11) Cost-effectiveness

BmEx = Biological monitoring of exposure
ExEx = Environmental monitoring of exposure
Verberk (1995)

pathways, they increase to overall marker response in an additive manner, but relative measures of association (e.g., relative risk, odds ratio, etc.) are based on the assumption of multiplicative association. As a result, the relative risk of response due to E_1, will be influenced by the background incidence response to E_2. In this situation, use of the risk difference rather than the relative risk to compare marker responses in persons exposed and unexposed to E_1 helps avoid the problem of dilution from a high background incidence from E_2" (Wilcosky, 1993). Studies to assess this type of statistical validity are difficult to accomplish because of the temporal factor, i.e., the time between the identification of the marker and the development of the disease. Identifying an early effect, i.e., an effect in pathogenesis or an effect predictive of disease, generally requires a prospective study design, although cross-sectional clinical or case–control studies of diseased and heavily exposed individuals can be used to great advantage. However, when not using a prospective design, care must be taken to avoid biased associations. This is often difficult and, hence, prospective studies are the best approach for validation. Prospective studies are expensive and time-consuming, and few are conducted. For example, despite the large number of studies on cytogenetic markers, there still has been little consensus on their predictive value, since most of the studies have been cross-sectional and suffer from temporal ambiguity.

Performing an appropriate prospective study, for example of cytogenetic markers, would take a large population and a relatively long time. A clever example of such a study is the Nordic prospective study on the relationship between peripheral lymphocyte chromosome damage and cancer morbidity in occupational groups (Brøgger et al., 1990; Hagmar et al., 1994). Ten laboratories in four Nordic countries participated in a study of a combined cohort of persons (mostly from occupational groups) who had been cyto-genetically tested. They found an association between chromosomal aberrations and cancer but not with sister-chromatid exchanges or micronuclei. In another prospective study in Italy, a similar finding was noted (Bonassi et al., 1995).

IV.5.1 Intervention studies

One type of prospective study that has been the focus of much interest is the intervention study or trial. Freedman & Schatzkin (1992) have assessed five types of validation questions using biomarkers (they call them intermediate end-points) in this context: (1) Does the intervention affect the biomarker? (2) Is the biomarker associated with prognostic or risk factors? (3) Is the biomarker associated with the main outcome? (4) Is the intervention effect on the main outcome, mediated by the biomarker? (5) Are the prognostic or risk factor effects mediated by the biomarkers? Freedman & Schatzkin (1992) show that each of these questions has different sample size requirements. The issues relating to intervention trials may also pertain to observational epidemiological studies (e.g., questions 2, 3 and 5).

IV.5.2   Attributable proportion

A measure of the degree of validation of an intermediate marker of effect is the extent to which the exposure is mediated through a marker. This may be assessed by calculating the attributable proportion, which has also been referred to in the literature as "population attributable risk" or "etiological fraction" (Benichou, 1991; Trock, 1995). The attributable proportion associated with a particular biomarker is an estimate of the proportion of diseased cases that must progress through the biomarker, i.e, the cases that would not occur if the event(s) resulting in the biomarker could be prevented (Schatzkin et al., 1990; Trock, 1995). The attributable proportion (AP) includes consideration of the sensitivity (S) of the assay and the relative risk (RR). It is defined as: $AP = S(1-(1/RR))$. The sensitivity is the factor with the greatest impact in the attributable proportion (Schatzkin et al., 1990). The attributable proportions takes into account both the strength of an association between a marker and disease and also the prevalence of the matter. Thus, for example, using data from a study by Perera et al. (1989) of DNA adducts in lung cancer cases and controls, Trock (1995) found that fewer than 50% of cases would have been attributable to the pathway, involving PAH adducts, suggesting that other pathways must be involved. Panels of multiple biomarkers might be better indicators of individuals at risk of cancer due to a particular exposure. Thus in the same study by Perera et al. (1989), when both PAH-DNA adducts and SCEs were examined together (individuals could be judged marker positive with either), the attributable proportion would be increased to more than 50%. Trock (1995) has described the next step. Once it has been established that a significant proportion of tumours can be attributed to a particular marker, it is necessary to examine the extent to which the marker truly represents a cellular event intervening between exposure and cancer. This can be evaluated using the epidemiological principles concerned with "intervening variables". In evaluating the relationship between an exposure and a disease outcome, one typically does not use statistical adjustment methods to adjust for a variable that is an intermediate step between exposure and outcome (Weinberg, 1993). Such an adjustment would sharply reduce or even eliminate the apparent effect of the exposure, since the marker's association with disease is a direct result of its association with exposure (assuming that the marker represents the relevant time period of exposure with respect to onset of disease) (Trock, 1995). One can take advantage of this property to assess the role of a marker as an intervening variable. If one compares the crude (i.e., unadjusted) RR for exposure to the RR for the exposure effect adjusted for the biological marker, the extent to which adjustment for the marker has reduced the apparent exposure effect indicates the degree to which the marker is linked to the exposure–disease relationship (Trock, 1995). If the effect of exposure occurs primarily through a pathway involving the marker, then the marker-adjusted exposure effect will essentially be eliminated, i.e., the adjusted RR will be close to 1.0 (Schatzkin et al., 1990).

IV.5.3   Predictive value

Another measure of validation of a biological marker of effect is the positive predictive value. Predictive value for a marker of disease is the proportion of people studied with a particular disease among all the people who have the marker. Predictive value is not only a property of the marker assay, it is determined by the sensitivity and specificity of the assay and the prevalence of the disease. Thus, for example, a marker that is 90% sensitive and 90% specific will still only have a predictive value of 50% when the prevalence of the underlying disease is 10%. Field studies that do not incorporate prevalence considerations in planning are unlikely to be able to detect an association between a marker of effect and disease, even if one exists (Schulte & Perera, 1993).

Positive predictive value and attributable proportion reflect very different things (Ottman, 1995; Khoury & Wagener, 1995). Positive predictive value, the risk of disease among persons with a specific marker, is important from the point of view of the individual. Attributable proportion, on the other hand, is the proportion of disease cases that must progress through the biomarkers and thus could be prevented if that process could be interrupted. This is important from an environmental or public health point of view.

IV.6. VALIDATION OF MARKERS OF SUSCEPTIBILITY

Exposure → Susceptibility → Disease

The tools of molecular biology and analytical chemistry have allowed researchers to identify a degree of inter-individual variability not previously imagined (Janetos, 1988). The biomarkers of this variability can be indicators of susceptibility to effects of exposure or to disease. As "effect modifiers" in epidemiological studies validate susceptibility biomarkers are useful and informative in portraying the nature and mechanism of a risk. Effect modification is a term with statistical and biological aspects. Statistically, the examination of joint effects of two or more factors is often discussed in the context of effect modification. It depends on the statistical method (e.g., multiplicative or additive) used to model interaction. From the biological perspective, effect modification conceptually answers the question of why two similarly exposed individuals do not develop a disease. The answer, in part, is individual variability in metabolic and detoxification capabilities (Schulte, 1993). The assessment of gene-environment interaction is important in the validation of biomarkers of susceptibility (and exposure). Gene-environment interaction is defined as "a different effect of an environmental exposure on disease risk in persons with different genotype," or alternatively, "a different effect of genotype on disease risk in persons with different environmental exposures" (Ottman, 1996). Ottman (1996) has described five biologically plausible models of gene–environment interaction, each of which leads to a different set of predictions about disease risk in individuals classified by presence or absence of a high-risk genotype or environmental exposure. These models are shown graphically in Fig. 16. If a biomarker of susceptibility is to be validated for disease, its relationship to both disease and exposure need to be determined.

Interactions between independent causal factors are inevitably confounded with dose–response and latency relationships. Dose–response refers to the changes in risk produced by changes in a single exposure, whereas interaction refers to changes in risk produced by two or more factors (e.g., an environmental factor and a genetic factor). Failure to adequately model dose–response and

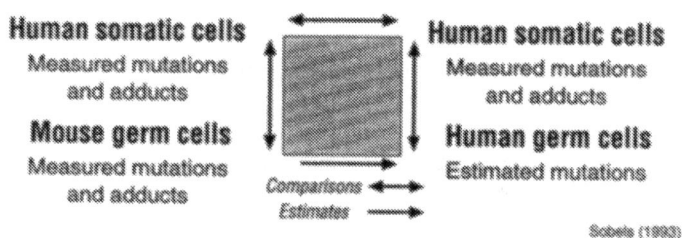

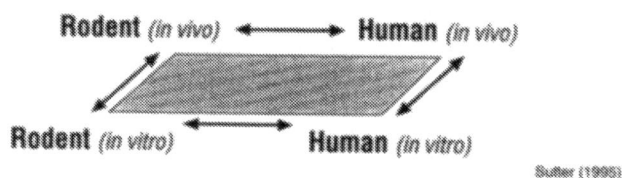

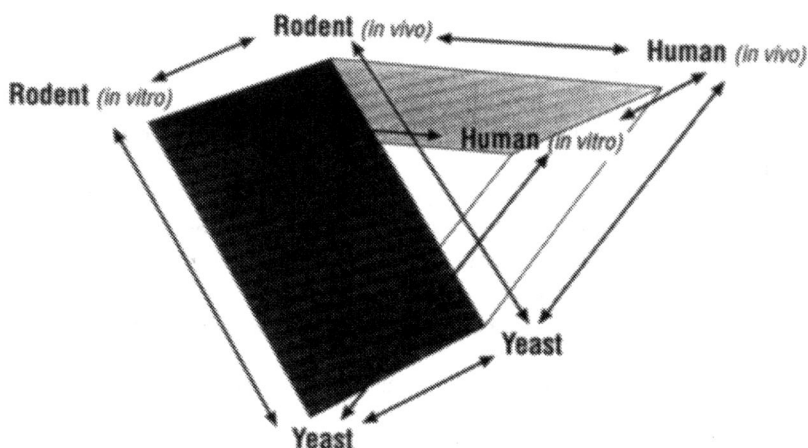

Fig. 16. Variations on the parallelogram approach

latency can lead to bias in interaction estimates (Greenland, 1993). Additionally, measurement errors, even if independent and non-differential, can distort interaction assessment. Since both genetic and environmental factors contribute to the etiology of most diseases, factors of each type are expected to confound or modify the effect of each other (Morganstern & Thomas, 1993).

The assessment of the role of susceptibility factors has a different set of issues depending on whether the focus is two factors that each have an independent causal effect of increasing the incidence of a particular disease or whether a factor has an effect on disease only when some other factor is present. The first instances are generally hereditary conditions, the second involves poly-morphisms in genes for metabolizing enzymes. These latter polymorphisms, which are of most interest in risk assessment, may not confer any risk of disease by themselves, but only when a particular exposure occurs (e.g., slow acetylators and aromatic amine exposure).

To validate a susceptibility marker it is important to minimize misclassification, which can occur by laboratory or epidemiological factors that affect phenotyping or genotyping (Rothman et al., 1993). A less-than-perfect assay, in terms of sensitivity and specificity, can bias the odds ratio in case-control studies. After minimizing misclassification, it is necessary to demonstrate that the marker either increases the biologically effective dose or elevates the risk of intermediate effects or disease.

Many of the epidemiological studies to validate markers of susceptibility have exhibited a high degree of heterogeneity (d'Errico et al., 1996). In a review of four genetically based metabolic polymorphisms involved in the metabolism of several carcinogens, d'Errico et al. (1996) identified a range of methodological features leading to discordant results. These include a high proportion of studies using prevalent cases, the frequent use of hospital controls, a low response rate, use of metabolic ratios as variables, and the lack of adequate adjustment for covariates. Additionally, such studies have been too small and had weak exposure characterization (Vineis, 1992).

In what may be viewed as a classic study, Vineis (1992) and colleagues showed an example of how partial validation of a susceptibility marker might occur without using disease as the outcome. Vineis (1992) compared the formation of haemoglobin adducts (which are documented surrogates for DNA adducts and are believed to be involved in carcinogenesis) between individuals exposed to 4-aminobiphenyl and who were slow or fast acetylators. They found that the slow acetylators had an average of 1.5-fold greater frequency of adducts than the fast acetylators. Despite these encouraging efforts at validation, few markers of susceptibility have been validated, and none are ready for use in population screening (Schulte & Halperin, 1987; OTA, 1990). However, rapid growth in interest in these types of markers have lead to the development of useful databases and the potential for incorporating them in risk assessments (Bois et al., 1995).

Khoury & Wagener (1995) have evaluated the relationships between genotype and exposure and found that the degree of increased disease risk in exposed susceptible individuals is highly dependent on the relationship between genotype and exposure. Information on this relationship is rarely available (Ottman, 1995).

Valid genetic tests could possibly improve the disease–predictive value of risk factors and establish a new paradigm in the primary prevention of many chronic diseases. This could be the identification and interruption of environmental cofactors that lead to clinical disease among persons with susceptibility genotypes (Khoury & Wagener, 1995).

IV.7. VALIDATION OF MULTIPLE BIOMARKERS

The use of multiple biomarkers has the potential to increase understanding of exposure, disease or susceptibility, but at the price of increasing the difficulty in combining the information from individual markers and interpreting the overall combination of markers. Perera et al. (1992) have demonstrated the richness of the information gained when they used a battery of biomarkers to assess genetic and molecular damage in residents from a polluted area of Poland (Perera et al., 1992). In this study, a common genotoxic model was used to provide a molecular link between environmental exposure and genetic alteration relevant to cancer and reproductive risk. Most of the biomarkers, carcinogen – DNA adducts, sister-chromatid exchanges, chromosomal aberrations, and *ras* oncogene over-expression were related to exposure to ambient PAH levels.

As larger batteries of biomarkers become available, the combinatorial strategies and techniques become more problematic. For example, any given nucleated cell type is generally estimated to contain 3000 to 8000 different proteins, while the total of all the different proteins required by all cell types during development and maturity is estimated to be between 30 000 and 50 000 (Anderson et al., 1981). Environmental exposures of cells is not likely to lead to perturbation of only one protein but rather many. Two-dimensional gel electrophoresis (2D-PAGE) has been used to detect small differences in protein composition in cells from subjects with and without various exposures. Although human observers are capable of searching large amounts of quantitative data for simple markers correlated with external information, global analysis, i.e., examination of the entire data set, for complex pattens of change is extremely difficult (Anderson et al., 1984). Environmental exposures are known to involve complex changes and thus require an approach capable of dealing with this type of data. Artificial intelligence and expert systems may provide a useful approach but they are not yet developed. Various image and multivariate analytical approaches have been useful. This is illustrated by the study of Robinson et al. (1995) who used 2D-PAGE to address the problem that there is no specific biochemical test for diagnosis of fetal alcohol syndrome (FAS). They tested serum proteins in 12 children with fetal alcohol syndrome and 8 sex- and age-matched children whose mothers did not consume alcohol. Multiple hypothesis testing on 34 of the gels consisting of more than 1700 spots per gel revealed 21 proteins that were classified as potential biomarkers of FAS on the basis of t-test (P < 0.02). No single protein differentiated all case subjects from control subjects, but a multivariate statistical procedure, stepwise canonical discriminate analysis, identified four spots that distinguished FAS case and control subjects with no misclassification.

The capacity to employ DNA arrays is revolutionizing our ability to analyse gene expression patterns, carry out genome-wide genetic mapping, clone members of gene families within and across species, scan for mutations in interesting genes, and define genes controlled by particular transcription factors (Blanchard & Hood, 1996; Borman, 1996). The expectation of researchers is that the expression of many genes will be measured before and after an exposure. This is the leading edge of efforts to use batteries of biomarkers to assess exposure, effect or susceptibility. The use of multiple biomarkers as a battery or in combination raises complex validation issues. The ability to make sense out of complex arrays of genetic expression data and multiple markers has little precedent. Much of the literature on the predictive value of medical tests usually concerns a single test with brief mention of multiple tests. Similarly, for multiple genes and environment interactions, there is little experience to address these situations. One approach that might be illustrative of handling large numbers of markers involves statistical analysis of protein arrays from 2D PAGE. Anderson et al. (1981, 1984) have described image analysis and multivariate statistical methods to analyse gene expression from two-dimensional gel electrophoresis. Using principal component analysis, and cluster analysis they demonstrated that a variety of human cultured cell types can be distinguished on the basis of complex patterns of expression of hundreds of proteins. This approach was also applied to assessing effects of xenobiotics (Anderson et al., 1981). The database from this allows for studying gene expression against the background of the real complexity of the cell.

Multivariate statistical analytical approaches are not without their problems. A review of the use of multivariate models in the general medical literature showed a general trend in overfitting data to models, no test for conformity of variables to a linear gradient, no mention of pertinent checks for proportional hazards, no reports on testing for interactions between independent variables, and unspecified coding and selection of independent variables (Concato et al., 1993). The investigators concluded that these problems make the results reported potentially inaccurate, misleading or difficult to interpret. These problems would be all the more evident in analysing large arrays of biomarkers such as DNA chips with tens of thousands of different types of DNA. In addition to analysing a gene for all heterozygous mutations, there still is a need to consider penetrance of the gene, phenotypic expression, data reduction and environmental covariates if this information is to be useful for environmental health research or risk assessment.

If multiple biomarkers are included in a multiple logistic regression model and one or more of these is measured with error, any of the effects may be under- or overestimated, even for exposure variables that are measured without error (Greenland, 1980; Kupper, 1984; Rosner et al., 1990; Greenland, 1993). Some of these problems are known to be manageable. Others will require environmental health researchers, statisticians, epidemiologists, risk assessors and laboratory scientists working with specialists in computer-driven "bioinformatics" (Rodbell, 1996). This is discussed in a later section.

IV.8. SOCIETAL IMPEDIMENTS TO VALIDATION

The greatest impediment to the validation of biomarkers has been the lack of awareness of the need for such studies and the funds to conduct them. There is often funding for laboratory development, but currently there is little research support for the scaling-up efforts and covariate assessments needed for population validation. This development should include modification of methods to reduce costs, which will be crucial for their use in large-scale population studies (Wogan, 1992).

Other impediments pertain to the ethical, legal and social issues. Many of these have been described (Schulte & Sweeney, 1995; Schulte et al., 1997). A key issue pertains to whether specimens collected for specific assays can be used for further, newer assays that may not have been developed or envisaged at the time the specimens were collected. This issue pertains to markers of exposure, effect and susceptibility, but mostly to susceptibility. The issues with susceptibility biomarkers include stigmatization of participants, prejudice and lack of opportunity due to having a particular genotype. In some countries, even acknowledging that a person had a genetic test can lead to these unwanted impacts. Identification of a particular genotypic polymorphism in a research participant may also have implications for the participant's family. The privacy and confidentiality pertaining to biomarker information is an important issue to research participants.

For scientists, the burden of recontacting participants of previous studies to obtain permission for further tests on the specimens can be too extensive to be worth doing. Moreover, it is likely that with stored specimens there will be the opportunity for numerous assessments that each could require repeated subject contact. There is also a range of opinions about the need to inform subjects of test results about markers with no known clinical significance. Currently, the balance between protecting subjects' rights and fostering efficient scientific exploration and investigation of factors that can affect public health is being debated.

IV.9. FRONTIERS OF VALIDATION

It would be a useful resource to characterize candidate markers of exposure, effect and susceptibility according to the extent of their validation. Two categories could be identified:

(1) Biomarkers with adequate assay development and laboratory validation – this would include determination of the limit of detection, sensitivity, specificity and reliability; and

(2) Biomarkers optimized for use in population studies:

   a)   optimization of specimen handling methods,
   b)   determination of biomarker frequency in population subgroups,
   c)   determination of confounding and effect modifying factors, and
   d)   determination of attributable proportion or predictive value.

Ideally, all candidate markers for various environmental health problems would be classified into these categories. To some extent, this has been accomplished for biomarkers of environmental exposure in a report by the IEH (IEH, 1996) in the United Kingdom and more generally in various National Research Council (NRC) documents (NRC, 1989a,b, 1992). For biomarkers of effect, this characterization would need to be done by disease or organ system as was begun in the NRC documents or some of the publications from the US National Cancer Institute. Biomarkers of susceptibility also have been chronicled in various journal articles. If there were such compendia of the validation of markers, researchers, funding agencies and environmental health organizations could consult these and use them in planning and funding decisions.

Two other resources can influence the extent to which biomarkers become validated. One is the ability to bank biological specimens or to access previously banked specimens. To the extent the specimens precede disease development in the participant donors and can be linked to disease outcomes, they serve as excellent opportunities to validate biomarkers and, in some instances, address some of the temporal issues. The other resource is also a bank, namely the bank of data that have resulted from various human and animal genome projects. Rodbell (1996) has suggested that environmental health researchers, utilizing appropriate computer selection strategies (bioinformatics) and toxicological insights, can select genes and gene products that play a role in environmental disease. Currently, much of the data in the genome data banks have not been validated in terms of their relationship to environmental disease.

Ultimately, the critical issue in discussing validation of biomarkers is to answer the question, "valid for what purpose?" Validation is a measure of degree, not an absolute determination. Any compendium of validation status will need revision and updating as new biomarkers are developed or as information about current biomarkers is enhanced. There is a need for critical thinking about when to use a biomarker in an epidemiological study instead of some more traditional measure of exposure or disease. Some of these criteria have been suggested (Rothman, 1993; Muscat, 1996). Too often, the temptation exists to use sensitive laboratory techniques to measure something merely because it can be measured rather than because it provides better and more useful information. This temptation should be resisted. Similarly in risk assessments the criteria for utilizing biological markers includes whether they add to the quality and believability of the risk assessment and whether they reduce uncertainty. Thus biomarkers that provide insight about mechanisms, support biological plausibility, or assist in refining risk estimates will be most useful.

ACKNOWLEDGMENTS

Marilyn Vine, Nat Rothman, and Eileen Kuempel are to be thanked for their comments on early drafts of this appendix and Karen Dragon for processing it.

IV.10. REFERENCES

Alessio L, Apostol P, & Crippa M (1995) Influence of individual factors and personal habits on the levels of biological indicators of exposure. Toxicol Letters, **77**: 93-103.

Armstrong BK, Whie E, & Saracci R (1994) Principles of exposure measurement in epidemiology. New York, Oxford University Press, pp 49-114.

Anderson NL, Taylor J, Scandora AE, Coulter BP, & Anderson NG (1981) The TYCHO system for computer analysis of two-dimensional gel electrophoresis patterns. Clin Chem, **27**: 1807-1820.

Anderson NL, Hoffman JP, Gemmell A, & Taylor J (1984) Global approaches to quantitative analysis of gene-expression patterns observed by use of two-dimensional gel electrophoresis. Clin Chem, **30**: 2031-2036.

Anderson S, Sadinski W, Shugart L, Brussard P, Depledge M, Ford T, Hose J, Stegemen J, Suk W, Wirgn I, & Wogan G (1994) Genetic and molecular ecotoxicology: a research framework. Environ Health Perspect, **102**(Suppl 12): 3-8.

Benichou J (1991) Methods of adjustment for estimating the attributable risk in case-control studies: a review. Statistics in Medicine, **IV**: 1753-1773.

Bernard AM (1995) Biokinetics and stability aspects of biomarkers: recommendations for application in population studies. Toxicol, **101**: 65-71.

Blanchard AP & Hood L (1996) Sequence to array: probing the genome's secrets. Nature Biotechnology, **14**: 1649.

Bois FY, Krowech G, & Zeise L (1995) Modeling human interindividual variability in metabolism and risk: the example of 4-aminobiphenyl. Risk Analysis, **15**: 205-213.

Bonassi S, Abbondandolo A, Camurri L, Dal Pra L, De Ferrari M, Degrassi F, Forni A, Lamberti L, Lando C, Padovani P, Sbrana I, Vecchio D, & Puntoni R (1995) Are chromosome aberrations in circulating lymphocytes predictive of future cancer onset in humans? Preliminary results of an Italian cohort study. Cancer Genet Cytogenet, **79**: 133-135.

Borman S (1996) DNA chips come of age. Chemical and Engineering News, December 9, pp 42-43.

Brøgger A, Hagmar L, Hansteen IL, Heim S, Hogstedt B, Knudsen L, Lambert B, Linnainmaa K, Mitelman F, Nordenson I, Reuterwall C, Salomaa S, Skerfving S, & Sorsa M (1990) An inter-Nordic prospective study on cytogenic endpoints and cancer risk. Cancer Genet Cytogenet, **44**: 85-92.

Concato J, Feinstern AR, & Holford TR (1993) The risk of determining risk with multivariate models. Ann Int Med, **118**: 201-210.

Denissenko MF, Pao A, Tang M, & Pfeifer GP (1996) Preferential formation of benzo(a)pyrene adducts at lung cancer mutational hotspots in p53. Science, **274**: 430-432.

D'Errico A, Taioli E, Chen X, & Vineis P (1996) Genetic metabolic polymorphisms and the risk of cancer: a review of the literature. Biomarkers, **1**: 149-173.

Ehrenberg L, Granath F, & Tornqvist M (1996) Macromolecular adducts as biomarkers of exposure to environmental mutagens in human populations. Environ Health Perspect, **104**(Suppl 3): 423-428.

Fennell TR, MacNeela JP, Thompson CL, & Bell DA (1996) Hemoglobin adducts from acrylonitrile and ethylene oxide in cigarette smokers: effects of glutathione transferase TI and MI genotypes. Toxicologist 30, 282 [Abstract No. 1443].

Freedman LS & Schatzkin A (1992) Sample size for studying intermediate endpoints within intervention trials of observational studies. Am J Epidemiol, **136**: 1148-59.

Galloway SM, Berry PK, Nickols WW, Wolman WR, Soper KA, Stolley PD, & Archer P (1986) Chromosome aberrations in individuals occupationally exposed to ethylene oxide and in a large control population. Mut Res, **170**: 55-57.

Gann P (1986) Use and misuses of existing data bases on environmental epidemiology: the case of air pollution. In: Kopfer FC & Craun GF (eds), Environmental Epidemiology, Lewis Publishers Inc., Chelsea, MI, pp 109-122.

Goyer RA & Rogan WJ (1986) When is biologic change an indicator of disease. In New and sensitive indicators of health impacts of environmental agents (Underhall DM & Radford ED (eds)). University of Pittsburgh, Pittsburgh, PA, pp 17-25.

Grandjean P, Brown SS, Reavey P, & Young DS (1995) Biomarkers in environmental toxicology: state of the art. Clin Chem, **41**: 1902-1909.

Greenland S (1980) The effect of misclassification in the presence of covariates. Am J Epidemiol, **112**: 564-569.

Greenland S (1993) Basic problems in interaction assessment. Environ Health Perspect, **101**(Suppl 4): 59-66.

Hagmar L, Brøgger A, Hansteen IL, Heim S, Hogstedt B, Knudson L, Lambert B, Linnainmaa K, Mitelman F, Nordenson I, Reuterwall C, Salomaa S, Skerfving S, & Sorsa M (1994) Cancer risk in humans predicted by increased levels of chromosomal aberrations in lymphocytes: Nordic study group on the health risk of chromosomal damage. Cancer Res, **54**: 2912-22.

Hatch M & Thomas D (1993) Measurement issues in environmental epidemiology. Environ Health Perspect, **101**(Suppl 4): 49-57.

Hattis D & Silver K (1993) The use of biomarkers in risk assessment. In: Schulte & Perera FP (eds). Molecular epidemiology: principles and practices, San Diego, Academic Press, pp 251-273.

Hernberg S & Aitio A (1987) Validation of biological monitoring tests. In: Occupational and environmental chemical hazards: cellular and biochemical indices for monitoring toxicity. Foa V, Emmetts EM, Maroni M & Columbi A (eds). Chichester, England, Ellis Horwood, pp 41-49.

Hulka BS (1991) Epidemiological studies using biological markers: Issues for epidemiologists. Cancer Epidemiology, Biomarkers and Prevention, **1**: 13-19.

Hulka BS & Margolin BH (1992) Methodologic issues in epidemiological studies using biomarkers. Am J Epidemiol, **135**: 200-204.

IEH (1996) The use of biomarkers in environmental risk assessment. Leicester, UK, Institute for Environment and Health.

Janetos AC (1988) Biological variability. In: Variations in susceptibility to inhaled pollutants, Brain JD, Beck BD, Warren AJ & Shaikh RA (eds). Baltimore, The John Hopkins University Press, pp 9-29.

Khoury MJ & Wagener DK (1995) Epidemiological evaluation of the use of genetics to improve the predictive value of disease risk factors. Am J Genet, **56**: 835-844.

Kriebel D (1994) The dosimetric model in occupational and environmental epidemiology. Occ Hyg, **1**: 55-68.

Kupper LL (1984) Effects of the use of unreliable surrogate variables on the validity of epidemiologic research studies. JASA, **72**: 481-488.

Louis TA (1988) General methods for analyzing repeated measures. Stat Med, **7**: 39-45.

Lucier GW & Thompson CL (1987) Issues in biochemical applications to risk assessment: When can lymphocytes be used as surrogate indicators. Environ Health Perspect, **76**: 189-191.

Margetts BM (1991) Basic issues in designing and interpreting epidemiological research. In: Margetts BM & Nelson M (eds) Design concepts in nutritional epidemiology. New York, Oxford University Press, pp 13-51.

McClellan RO (1995) Risk assessment and biological mechanisms: lessons learned, future opportunities. Toxicology, **102**: 239-258.

Morgenstern H & Thomas D (1993) Principles of study design in environmental epidemiology. Environ Health Perspect, **101**(Suppl 4): 23-28.

Muscat JE (1996) Epidemiological reasoning and biological rationale. Biomarkers, **1**: 144-145.

NRC (National Research Council) (1987) Biological marker in environmental health research. Environ Health Perspect, **74**: 1-19.

NRC (National Research Council) (1989a) Biological markers in reproductive toxicology. Washington, DC, National Academy Press.

NRC (National Research Council) (1989b) Biological markers in pulmonary toxicology. Washington, DC, National Academy Press.

NRC (National Research Council) (1992) Biological markers in immunotoxicology. Washington, DC, National Academy Press.

OTA (Office of Technology Assessment) (1990) Genetic monitoring and screening in the workplace. US Congress, OTA, Washington, DC, US Government Printing Office.

Olden K & Klein JL (1995) Environmental health science research and human risk assessment. Mol Carcin, **14**: 2-9.

Omenn GS (1995) Assessing the risk assessment paradigm. Toxicolog, **102**: 23-28.

Ottman R (1995) Gene-environment interactions and public health. Am J Genet, **56**: 821-823.

Ottman R (1996) Gene-environment interaction: definition and study designs. Prev Med, **25**: 764-770.

Pearce N, de Sanjose S, Boffetta P, Kogevinas M, Saracci R, & Savitz D (1995) Limitations of biomarkers of exposure in cancer epidemiology. Epidemiology, **6**: 190-193.

Perera F (1987) The potential usefulness of biological markers in risk assessment. Environ Health Perspect, **79**: 141-145.

Perera FP (1996) Molecular epidemiology: insights into cancer susceptibility, risk assessment and prevention. J Natl Cancer Inst, **88**: 496-509.

Perera F, Mayer J, Jaretzki A, Hearne S, Brenner D, Young TL, Fishman H, & Grimes M (1989) Comparison of DNA adducts and sister chromatid exchanges in lung cancer cases and controls. Cancer Red, **49**: 4446-4451.

Perera FP, Hemminki K, Gryzbowska E, Motyklewicz G, Michalska J, Santella RM, Young TL, Dickey C, Brandt-Rauf P, DeVivo I, Blaner W, Tsai WY, & Chorazy M (1992) Molecular and genetic damage in humans from environmental pollution in Poland. Nature, **360**: 254-258.

Rhomberg L (1995) Estimation and evaluation of dose. In: Farland W, Olin S, Park C, Rhomberg L, Scheupien R, Starr T, & Wilson J (eds). Low-dose extrapolation of cancer risks: issues and perspectives. Washington, DC, International Life Sciences Institute Press, pp 61-74.

Robinson MK, Myrick JE, Henderson LO, Coles CP, Powell MK, Orr GA, & Lemkin PF (1995) Two-dimensional protein electrophoresis and multiple hypothesis testing to detect potential serum biomarkers in children with fetal alcohol syndrome. Electrophoresis, **16**: 1176-1183.

Rodbell M (1996) Bioinformatics: an emerging means of assessing environmental health. Environ Health Perspect, **104**: 136.

Rosner B, Spiegelman D, & Willett WC (1990) Correction of logistic regression relative risk estimates and confidence intervals for measurement error: the case of multiple covariates measured with error. Am J Epidemiol, **132**: 734-745.

Ross R, Yuan JM, You MC, Wogan GN, Ian GS, TU J, Groopman JD, Gao YT, & Henderson BE (1992) Urinary aflatoxin biomarkers and risk of hepatocellular carcinoma. Lancet, **339**: 943-946.

Rothman N (1993) Epilogue. In: Molecular epidemiology: principles and practices. Schulte PA & Perera FP (eds). San Diego, Academic Press.

Rothman N, Stewart WF, Caporaso NE, & Hayes RB (1993) Misclassification of genetics susceptibility biomarkers: implications for case-control and cross population comparison. Cancer Epid Biomarkers, Prevention, **2**: 299-303.

Rothman N, Stewart WF, & Schulte PA (1995) Incorporating biomarkers into cancer epidemiology: a matrix of biomarker and study design categories: Cancer Epid Biomarkers, Prevention, **4**: 301-311.

Schatzkin A, Freedman LS, Schiffman MH, & Dawsey J (1990) Validation of intermediate endpoints in cancer research. J Natl Cancer Inst, **82**: 1746-1752.

Schulte PA (1989) A conceptual framework for the validation and use of biomarkers. Environ Res, **48**: 129-144.

Schulte PA (1993) Use of biological markers in occupational health research and practices. J Tox Env Health, **40**: 359-366.

Schulte PA & Halperin WE (1987) Genetic screening and monitoring of workers. In: Recent advances in occupational health. Edinburgh, Churchill Livingstone, **3**: 135-154.

Schulte PA & Mazzuckelli LF (1991) Validation of biological markers for quantitative risk assessment. Environ Health Perspect, **90**: 239-246.

Schulte PA & Perera FP (1993) Validation. In: Schulte PA & Perera FP (eds), Molecular epidemiology: principles and practices, San Diego, CA, Academic Press, pp 79-107.

Schulte PA & Sweeney MH (1995) Ethical considerations, confidentiality issues, rights of human subjects, and uses of monitoring data in research and regulation. Environ Health Perspect, (Suppl 3): 69-74.

Schulte PA & Talaska G (1995) Validity criteria for use of biological markers of exposure to chemical agents in environmental epidemiology. Toxicol, **101**: 73-88.

Schulte PA & Waters M (1999) Using molecular epidemiology in assessing exposure for risk assessment. Ann NY Acad Sci, **895**: 101-111.

Schulte PA, Boeniger M, Walker J, Schober SE, Pereira MA, Gulati DK, Wojciechowski JP, Garza A, Froelich R, Strauss G, Halperin WE, Herrick R, & Griffith J (1992) Biological markers in hospital workers exposed to low levels of ethylene oxide. Mutat Res, **278**: 237-251.

Schulte PA, Rothman N, & Schottenfield D (1993) Design consideration in molecular epidemiology. In: Schulte PA & Perera FP (eds), Molecular epidemiology: principles and practices, San Diego, CA, Academic Press, pp 159-198.

Schulte PA, Hunter D, & Rothman N (1997) Ethical and social issues in the use of biomarkers in epidemiological research. In: Toniolo P, Boffetta P, Shuker DEG, Rothman N, Hulka B, & Pearce N (eds) Application of biomarkers in cancer epidemiology. IARC Scientific Publication No. 142. Lyon, International Agency for Research on Cancer, pp 313-318.

Sexton K, Reiter LW, & Zenick H (1995) Research to strengthen the scientific basis for health risk assessment: a survey for the context and rationale for mechanistically based methods and models. Toxicol, **102**: 3-20.

Shugart LR, McCarthy JF, & Halbrook RS (1992) Biological markers of environmental and ecological contamination: an overview. Risk Analysis, **12**: 353-360.

Sobels FH (1993) Approaches to assessing genetic risks from exposure to chemicals. Environ Health Perspect, **101**(Suppl 3): 327-332.

Stevens DK, Bull RJ, Nauman CH, & Blancato JN (1991) Decision model for biomarkers of exposure. Regulatory Toxicol Pharmacol, **14**: 286-296.

Sutter JR (1995) Molecular and cellular approaches to extrapolation for risk assessment. Environ Health Perspect, **103**: 386-389.

Trock BJ (1995) Application of biological markers in cancer environmental epidemiology. Toxicol, **101**: 93-98.

Verberk MM (1995) Biomarkers of exposure versus parameters of external exposure; practical applications in estimating health risks. Toxicol, **101**:107-115.

Vine MF (1996) Biologic markers of exposure: current status and future research needs. Toxicol Ind Health, **12**: 189-200.

Vineis P (1992) Uses of biochemical and biological markers in occupational epidemiology. Rev Epidem Sante Publ, **40**: 563-569.

Vineis P & Porta M (1996) Causal thinking, biomarkers, and mechanisms of carcinogenesis. Clin Epidemiol, **49**: 951-956.

Ward JB Jr & Henderson RE (1996) Identification of needs in biomarker research. Environ Health Perspect, **104**(Suppl 5): 895-900.

Weinberg CR (1993) Toward a clearer definition of confounding. Am J Epidemiol, **137**: 1-8.

White E (1997) Effects of biomarker measurement error on epidemiologic studies. In: Application of biomarkers in cancer epidemiology. Toniolo P (eds) IARC Scientific Publication No. 142, Lyon, International Agency for Research on Cancer, pp 73-93.

Wilcosky TC (1993) Biological markers of intermediate outcomes in studies of indoor air and other complex mixtures. Environ Health Perspect, **101**(Suppl 4): 193-197.

WHO (1975) Early detection of health impairments in occupational exposure to health hazards. WHO Technical Report 571, Geneva, World Health Organization.

Wogan GN (1992) Molecular epidemiology in cancer risk assessment and prevention: recent progress and avenues for future research. Environ Health Perspect, **98**: 167-178.

RÉSUMÉ ET CONCLUSIONS

1. Résumé

Le groupe spécial sur les biomarqueurs des critères d'hygiène de l'environnement, s'appuyant sur les catégories et évaluations antérieures des biomarqueurs de la recherche, les a étudiés dans le cadre de l'évaluation du risque. On a défini cette dernière comme étant la série d'étapes séparant la recherche de la gestion du risque. Elle fournit à la société des estimations du risque lorsqu'il existe une incertitude quant à la sécurité des degrés d'exposition courants ou futurs à des substances toxiques présentes dans l'environnement et dans les milieux professionnels.

Le groupe spécial a élaboré un cadre de référence permettant de choisir et de valider les biomarqueurs dont on a évalué le risque. Des exemples ont été cités concernant la manière dont les trois types de biomarqueurs, d'exposition, d'effet et de sensibilité pouvaient être validés pour la recherche et employés dans les évaluations du risque. Des biomarqueurs validés peuvent permettre des évaluations du risque ayant un fondement biologique.

Il existe peu d'exemples dans lesquels les biomarqueurs validés ont été employés pour des évaluations quantitatives du risque. Les travaux futurs devront comprendre des actions sur le plan scientifique, technique, organisationnel et administratif afin de coordonner les efforts visant à fixer un programme pour la recherche sur les biomarqueurs qui permettra de mener à bien les évaluations des risques importants. Il faudra pour cela s'engager dans une collaboration à long terme et mener des études prospectives afin de relier les biomarqueurs aux risques de maladies.

2. Conclusions

Les biomarqueurs validés servent à réduire l'incertitude dans les évaluations du risque. Toutefois, ces biomarqueurs doivent être considérés comme une autre série d'outils à la disposition des chercheurs et de ceux qui évaluent les risques, et non pas se substituer aux approches traditionnelles.

La validation des biomarqueurs pour la recherche et l'évaluation du risque exige des études de laboratoire et des études épidémiologiques.

Utiliser avec succès les données relatives aux biomarqueurs suppose que l'on comprenne le mécanisme. Il est certainement important d'incorporer des données mécanistes, mais les évaluations du risque et les réglementations ne doivent pas attendre l'élaboration de ces données, pas plus que l'incertitude quant aux mécanismes en jeu ne peuvent être utilisés pour bloquer l'action de santé publique.

L'apport des biomarqueurs de la sensibilité renferme un potentiel important, mais qui devra être réalisé à grande échelle dans l'évolution quantitative du risque.

Un engagement à long terme vis-à-vis de l'évaluation de la validité des biomarqueurs pour l'estimation du risque, la recherche sur l'hygiène du milieu et les pratiques de santé publique est aujourd'hui nécessaire.

RESUMEN Y CONCLUSIONES

1. Resumen

El Grupo Especial sobre Criterios de Salud Ambiental para los Biomarcadores, a partir de categorizaciones y evaluaciones anteriores de biomarcadores para fines de investigación, los examinó para evaluar los riesgos. La evaluación de riesgos se definió como la serie de medidas que deben adoptarse entre la investigación y la gestión de los riesgos. La sociedad obtiene así estimaciones del riesgo cuando existe incertidumbre en cuanto a la inocuidad de los niveles de exposición, presentes o futuros, a las sustancias tóxicas medioambientales y ocupacionales.

El Grupo Especial elaboró un marco para seleccionar y validar biomarcadores con fines de evaluación de riesgos. Se citaron ejemplos sobre cómo podrían validarse tres tipos de biomarcadores - de la exposición, de los efectos y de la susceptibilidad - con fines de investigación y como instrumentos de evaluación de los riesgos. Los biomarcadores válidos permitirían realizar evaluaciones de riesgos biológicamente fundamentadas.

Son pocos los casos de biomarcadores validados que se hayan empleado en evaluaciones cuantitativas de los riesgos. Como parte de los futuros trabajos deberían llevarse a cabo actividades científicas, técnicas, organizacionales y administrativas para coordinar los esfuerzos encaminados a establecer un programa de investigaciones sobre biomarcadores que faciliten la realización de evaluaciones de riesgos importantes. Esto requerirá compromisos de colaboración a largo plazo y la realización de estudios prospectivos para establecer la relación entre los biomarcadores y los riesgos de enfermedad.

2. Conclusiones

Los biomarcadores validados son útiles para reducir la incertidumbre en las evaluaciones de riesgos. Sin embargo, deben considerarse como un instrumento más a disposición de los investigadores y los evaluadores de riesgos, no como una alternativa para reemplazar la manera tradicional de proceder.

La validación de biomarcadores para fines de investigación y de evaluación de riesgos requiere la realización de estudios tanto de laboratorio como epidemiológicos.

Para utilizar de modo satisfactorio los datos sobre biomarcadores es preciso comprender el mecanismo implicado. La incorporación de datos mecanísticos en la evaluación de los riesgos es sin duda importante, pero esa evaluación y su reglamentación deben hacerse sin esperar a disponer de datos mecanísticos, y no deberá bloquearse la acción de salud pública alegando que existe incertidumbre sobre el mecanismo.

Los biomarcadores de la susceptibilidad encierran un potencial considerable, que no obstante tiene que confirmarse mediante la realización de evaluaciones cuantitativas de riesgos en gran escala.

Se necesita un compromiso a largo plazo en lo que respecta a la evaluación de la validez de los biomarcadores para la evaluación de riesgos, para las investigaciones de salud ambiental y para la práctica de salud pública.

www.ingramcontent.com/pod-product-compliance
Lightning Source LLC
Chambersburg PA
CBHW051716020426
42333CB00014B/1010